The Role of Calcium and Comparable Cations in Animal Behaviour

Dedication
We dedicate this to the memory of our parents.

The Role of Calcium and Comparable Cations in Animal Behaviour

by
R. G. Wilkins and P. C. Wilkins
New Mexico State University, Las Cruces, New Mexico, USA
e-mail: pwilkins@nmsu.edu

advancing the chemical sciences

ISBN 0-85404-666-6

A catalogue record for this book is available from the British Library

Published by The Royal Society of Chemistry,
Thomas Graham House, Science Park, Milton Road, Cambridge CB4 0WF, UK

Registered Charity Number 207890
For further information see our web site at www.rsc.org

Typeset by Keytec Typesetting Ltd
Printed and bound by TJ International Ltd, Padstow, Cornwall, UK

Preface

The biologically essential s-group elements sodium, potassium, magnesium and calcium in the form of their ions are colourless and show no redox behaviour. They have therefore been largely ignored by the spectroscopically-minded bioinorganic fraternity and mainly taken for granted by the physiologists. All however appreciate their importance. In this book we link the behaviour, sometimes anomalous, of creatures both great and small with some aspect of these elements. Calcium features heavily in the book and justifies its specific inclusion in the title. It is of course no more essential than the comparable cations, namely sodium, potassium and magnesium, but it has many more and varied roles to play in the functions of an animal. Calcium is also often part of the solid structures that are so vital. We also include quite a bit about the myriad of channels that allow the ions to move across membranes and thereby carry out their various tasks.

It was neither a requirement nor our intention to go into the details that are found in the appropriate texts or reviews. We have also kept the references to a minimum, concentrating on those that will give the reader entry to the vast literature, of which we have only scratched the surface. We do hope however that we have provided sufficient background so that the reader can understand how these ions work and the many examples of them in action.

This has been a fun book to write. It has not been easy because our own training and research interests have involved inorganic reaction mechanisms and iron bioinorganic chemistry. We have therefore enjoyed learning (hopefully correctly) many new things about the workings of creatures, all the time focusing on the contributions of our s-group elements. Where would life be without them?

Thanks go to Dr Michael Johnson and a reviewer who made very helpful comments. Any errors that remain are of course entirely our own. Thanks also to Mrs Gail Freel who helped with the clip-art pictures. Mrs

Janet Freshwater of the Royal Society of Chemistry was encouraging and helpful at all stages. We appreciate all the help we received from Katrina Turner and the unsung specialists at the RSC. Again, thank you to all these people.

R.G.W.   Las Cruces, NM, USA
P.C.W.   September, 2002

Contents

Contents

Glossary

abductor: A muscle that moves a limb away from the middle of the body of an animal. It works antagonistically with adductor muscles.

action potential: A transient change in the electrical potential across an excitable membrane of a nerve or muscle cell. The movement of Na^+, K^+, Ca^{2+} and Cl^- ions is involved.

active zone: A region in a presynaptic nerve terminal that is believed to be the site of transmitter release.

adductor: A muscle that pulls a limb inwards.

afferent: Conducting towards or inwards.

agonist: A substance that produces the same response in a receptor as does the endogenous chemical.

Amphibia: A class of vertebrate chordates that includes frogs, toads, newts and salamanders.

anoxia: The absence or reduced amounts of oxygen in tissues.

antagonist: A chemical that competes with another (agonist) and inhibits some cell response. It can be competitive (binding at the same site) or noncompetitive (binding at a different site).

apical (surface): The surface of an epithelial cell that faces the lumen.

aorta: The main artery through which blood flows from the heart to the rest of the body.

apoptosis: Natural or programmed cell death that is required for normal animal development.

Arabidopsis: Genus of flowering plants.

arteriosclerosis: The thickening (hardening) of artery walls.

Arthropoda: Invertebrates with segmented bodies, jointed limbs and external skeletons.

Articulata: A class of brachiopod with shells that are joined by a hinge joint comprising two teeth on one part that move in sockets on the other part.

astrocytes: One class of glia cells. Oligodendrocytes are the other main type of glia cells in the CNS.

ataxia: Some loss of control over bodily movements.

ATPase: Any enzyme that catalyses the hydrolysis of ATP or the synthesis of ATP (ATP synthetase).

atrium: One of the two upper cavities in the heart where blood is received from veins.

autonomic nervous system: That part of the nervous system that innervates involuntary (unconsciously controlled) internal organs. There are two divisions, the parasympathetic and the sympathetic.

axon: The long part of a nerve cell that conducts impulses from the soma.

basolateral (surface): The surface of an epithelial cell that adjoins underlying tissue.

biogenic: Produced by a living organism.

botulinus toxin: The most toxic substance known to man and often the cause of food poisoning.

bradycardia: Abnormally slow heart rate (<60 beats min^{-1}), opposite to tachycardia.

buccal ganglion: A mass of nerve cell bodies in the mouth.

calcisome: Calcium store in non-muscle cells.

calculus: Minerals, usually as stones, that form within the body.

carapace: The hard upper shell of a crustacean.

carotid body: Tissues adjacent to the carotid sinus with chemoreceptors that are sensitive to $[O_2]$, $[CO_2]$ and pH in the blood.

carotid sinus: Occurs at a branch of the carotid artery (about an inch above the collar bone), the blood vessel that begins at the aorta and runs straight up through the neck. Contains pressure receptors to gauge blood pressure.

central nervous system, CNS: In vertebrates these are a collection of neurons contained in the brain and spinal cord that process information to and from the peripheral nervous system (PNS).

cephalopods: An advanced class of molluscs that includes squid, cuttlefish and octopus.

channelopathies: Defects in ion channel functions that cause diseases in mammals.

Charcot–Marie–Tooth (CMT) disease: Results in progressive peripheral nerve degeneration leading to weakness in arms and legs.

chitin: A polymer of *N*-acetylglucosamine that is found in the exoskeletons of arthropods and fungi.

cholinergic: A nerve fibre that releases ACh when stimulated or one that is stimulated by ACh.

cilium: Short hairlike structure covering many cells and that beat in a coordinated back-and-forth motion producing cell movement.

coccolithophorids: Unicellular photosynthetic organisms.

cochlea: Small snail shell-shaped bone found in the inner ears of mammals.

collagen: An insoluble fibrous protein found in the connective tissues of skin, tendons and bones.

congenital: A defect or disease that has existed from birth.

copepods: The minute components of plankton.

cortex: The outer layer of a body organ immediately below the outer membrane.

crustacean: An arthropod with a hard shell and that is usually aquatic.

cytokinesis: One of the stages in cell division, *i.e.* cytoplasmic division into two daughter cells.

cytolysis: The destruction of the outer membrane resulting in the breakdown of the cell.

cytoplasm: All the material within the cell except the nucleus.

cytosol: All the material within the cell except the nucleus and membrane-bound organelles.

dendrite: A branched process of a nerve cell.

diatom: Microscopic unicellular alga that is found as plankton and that forms fossil deposits.

dinoflagellates: A group of unicellular protoctista that are the main constituents of marine and freshwater plankton.

diuretic: A compound that increases urination.

dorsal: At or near the back.

efferent: Carrying away (from body centre).

electroencephalograph: An instrument that receives and records electric impulses produced by brain cells.

electroolfactograph: An instrument that records electrical impulses produced by odours.

endocrine: Ductless glands that synthesize hormones and secrete them directly into the blood or other intercellular fluids.

endocytosis: Incorporation of material from the cell exterior into vesicles inside the cell.

endogenous: Originating within the organism, tissue or cell.

endolymph: The fluid in the membranes of the ear.

endoplasmic reticulum (ER): A system of continuous convoluted membranes in the cytoplasm of eukaryotic cells and is the site of protein synthesis.

endorphins: Compounds in the brain that can effect pain relief.

endoskeleton: Internal skeleton, the supporting framework that lies entirely within the body of an animal.

endothelium: A single layer of cells that line the internal surfaces of the heart, blood and lymphatic vessels.

epithelium: A sheet of cells that covers the outer surfaces of the body and the walls of internal cavities in multicellular organisms.

eukaryote: All organisms (except bacteria) that contain cells in which genetic material is contained within the nucleus.

excitable cells: Living cells or tissue that respond rapidly to stimuli.

excitation-contraction (EC) coupling: Stimulation of a neuron that results in the contraction of a muscle.

exocrine: Glands that secrete specialized substances through a duct into a body cavity (*e.g.* gut) or body surface (skin).

exocytosis: Release of material from synaptic vesicles to the exterior of eukaryotic cells.

exoskeleton: A rigid external covering that protects and supports the body it encases.

extrafusal: Muscle fibres outside the muscle spindle organ. They make up the bulk of skeletal muscle.

febrile: High body temperature.

flaccid: Weak and flabby, lacking muscle tone (balanced muscle tension) accompanied by decreased reflex response and loss of muscle bulk.

flagellum: Flexible outgrowth from a cell surface, similar to but much longer than a cilium in being able to produce motion (*e.g.* of sperm).

foraminifer: Protozoa having perforated shell through which pseudopods emerge.

fibre: A threadlike aggregation of cells in animals forming a strand.

fibril: Component part of a fibre.

ganglion: A bunch of nerve cells enclosed in a sheath outside the CNS in vertebrates. In invertebrates they occur along nerve cords and are part of the nervous system.

gastropod: Molluscs that move by using a large muscular foot, *e.g.* snail, limpet.

glia (= neuroglia): Several subclasses of connective tissue that support the CNS, but that are now also believed to have further roles possibly including signalling.

G-proteins: Guanine-nucleotide binding proteins in the cytoplasmic faces of the plasma membranes of mammalian cells. They relay signals from certain hormone and neurotransmitter receptors to intracellular pathways.

haemolymph: The colourless fluid containing white blood cells that is conveyed through the body by the lymphatic system.

hippocampus: A part of the vertebrate brain the function of which is related to the expression of emotions like fear and anger.

homeostasis: The essential regulation by an organism of the composition of its internal environment.

ionotropic: A receptor that mediates its effects by regulating ion channels.

invertebrate: An animal without a backbone.

ischemia: Reduced blood supply to an organ resulting in lack of oxygen, tissue death and localized death.

lateral line: A visible canal just below the surface of the skin, which runs down each side of a fish and that has numerous openings to the external environment.

limnology: The study of lakes and other freshwater systems (standing water).

lumen: The interior of a cavity or duct such as a blood vessel, gut or subcellular organelle, *e.g.* SR.

lymph: The fluid bathing all tissues and that is in contact with blood in lymphoid tissues.

marine colloids: The most abundant particles in the ocean that range in size from about 1 nm to 1 μm in diameter.

metabotropic: A receptor that mediates its effects by the activation of enzymes.

metazoa: All multicellular animals except sponges.

mollusca: An invertebrate with a soft body and usually a hard outer shell.

muscle spindle: A stretch receptor in vertebrate muscles that is localized between *extrafusal* muscle fibres.

multipolar neuron: A neuron with one axon and several major processes (branching into dendrites) extending in different directions from the cell body.

myasthenia: A condition that causes weakness in certain muscles.

myoclonus: Muscle spasms with alternating contractions and relaxations.

myoplasm: The cytosol in a muscle cell.

myocyte: A muscle cell.

myotonia: Any condition in which muscles do not relax after contraction. Also, an hereditary disease associated with specific muscle malformations.

necrosis: Localized tissue death.

nematode: A species of roundworm.

nerve fibre: The slender thread (axon) of a neuron and associated tissues.

neuropath: A person affected by a nervous disease.

nociceptor: A receptor that reacts to pain.

opioid receptor: A site on a cell surface that interacts specifically with an opioid (narcotic).

organelle: A structure within a *eukaryotic* cell that has a specific function.

osmolality: The effective pressure that is required to prevent osmotic flow.

osmosis: The movement of water through a membrane that is permeable to water.

osteoporosis: Brittle bones resulting from the loss of bony tissue.

otolith: Small particles of $CaCO_3$ in the inner ear.

pacemaker cell: The sinuatrial node in the heart that initiates and maintains the normal heartbeat.

paramecium: A slipper-shaped freshwater protozoan covered with cilia.

parasympathetic nervous system: A division of the *autonomic nervous system*, which leaves the cranial and sacral segments of the CNS. Generally controls resting body functions.

parietal cells: Cells in the wall of the stomach that produce HCl in the gastric juice secreted by glands in the stomach lining.

pathology: The science of bodily diseases; also the symptoms of a disease.

perineurium: The connective tissue sheath around a bundle of nerve fibres within a nerve.

peripheral nervous system (PNS): The nerves outside the brain and spinal cord.

pit vipers: Any US snake of the *Crotalidae* family with a pit between the eye and nostril.

plasmalemma (also plasma membrane or cell membrane): The membrane that forms a semi-permeable boundary between the cell and subcellular compartments.

plasmodium: A parasitic protozoan of the genus *Plasmodium*, including those that cause malaria.

polymerase chain reaction (PCR): A technique used to replicate a DNA fragment and produce many copies of a particular DNA sequence.

process: A natural outgrowth from a structure, *e.g.* the axon or dendrite of a nerve cell.

prokaryote: Bacteria and viruses that contain genetic material not enclosed in a cell nucleus.

Protoctista: Unicellular or simple multicellular organisms with nuclei, but that are not animals, plants or fungi.

resting potential: The electrical potential across the membrane of a nerve or muscle cell at rest.

ribozyme: An RNA molecule that can catalyse internal changes.

saltatory conduction: Impulse jumping from node to node in myelinated nerve fibres.

sarcolemma: The contractile membrane sheath around a muscle fibre.

sarcoplasmic reticulum (SR): The specialized form of the endoplasmic reticulum (ER) that is a network of membrane-lined cavities surrounding myofibrils and acts as a Ca^{2+} ion store.

second messenger: Intracellular regulatory agents that are controlled by an extracellular first messenger.

second-order neuron: A neuron that receives input from primary sensory neurons.

serum: The amber coloured fluid component of blood that remains after removal of fibrin and clotting agents.

somatic: Pertaining to the organs and tissues of the body other than the gut and associated structures.

sympathetic nervous system: A division of the *autonomic nervous system* that arises from the thoracic and lumbar segments of the CNS. Generally controls stress.

synapse: The junction of two nerve cells.

tachycardia: A regular, but abnormally high ($>$100 beats min^{-1}) heartbeat.

teleost: Includes most modern bony fishes with body skeletons that usually also have a swimbladder or lung.

tetanus: Prolonged muscle contraction caused by rapidly repeated motor impulses.

tight junction: Membrane fusion, under the influence of Ca^{2+} ions, between adjacent cells that prevents extracellular material from passing between the adjoining cells.

transduction: Generally refers to the modulation of one kind of energy by another.

umbrella: The gelatinous disc of, for example, a jellyfish that expands and contracts to propel the creature through water.

vacuole: A large membrane-bound cavity in the cytoplasm of a eukaryotic cell.

ventricle: A cavity or chamber, *e.g.* in the heart, which receives blood from the aorta and pumps it into the arterial system.

vertebrates: Animals that posses a backbone, a brain within a skull, a circulatory system, a heart, ears, kidneys and other organs.

vesicle: A small sac confined within a membrane and in the cytoplasm of a cell.

vestibule: A space serving as an entrance to a passageway.

visceral: Relating to the internal organs in the body cavities of animals.

Amino Acids and their Abbreviations

Structure	Name	3-letter symbol	1-letter symbol

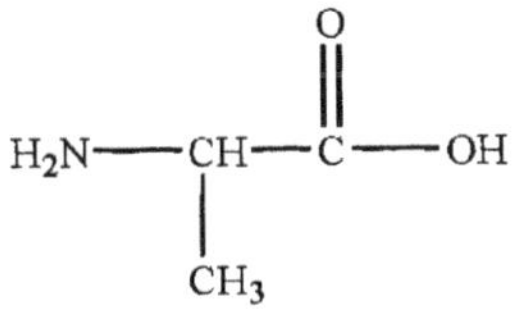

$$H_2N-\overset{\displaystyle |}{\underset{\displaystyle CH_3}{CH}}-\overset{\displaystyle O}{\overset{\displaystyle \|}{C}}-OH$$

alanine — ala — A

$$H_2N-\overset{\displaystyle |}{CH}-\overset{\displaystyle O}{\overset{\displaystyle \|}{C}}-OH$$

CH$_2$

CH$_2$

CH$_2$

NH

C$=$NH

NH$_2$

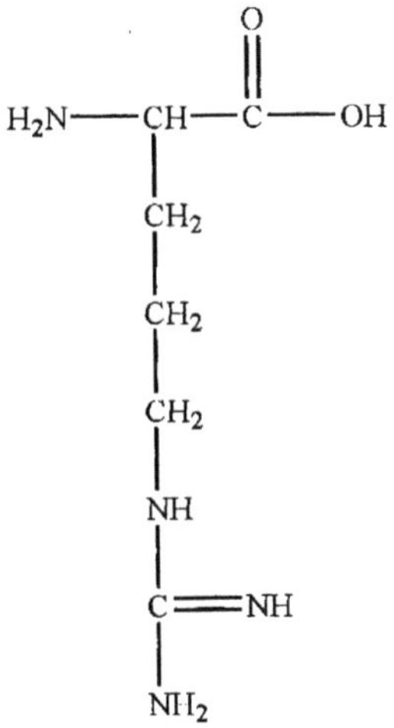

arginine — arg — R

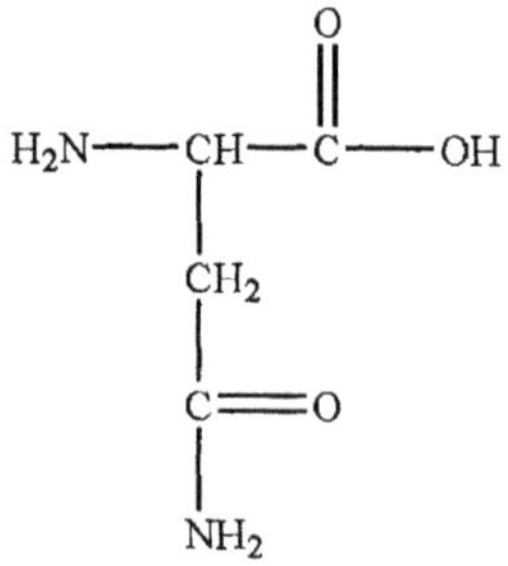

asparagine   asn   N

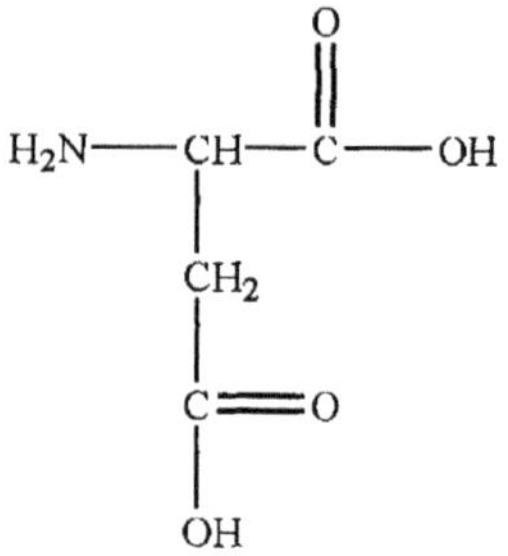

aspartic acid   asp   D

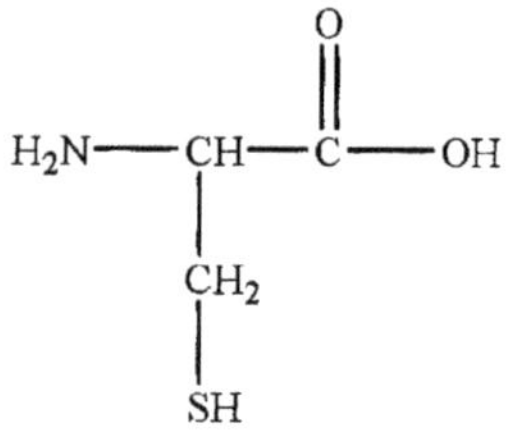

cysteine   cys   C

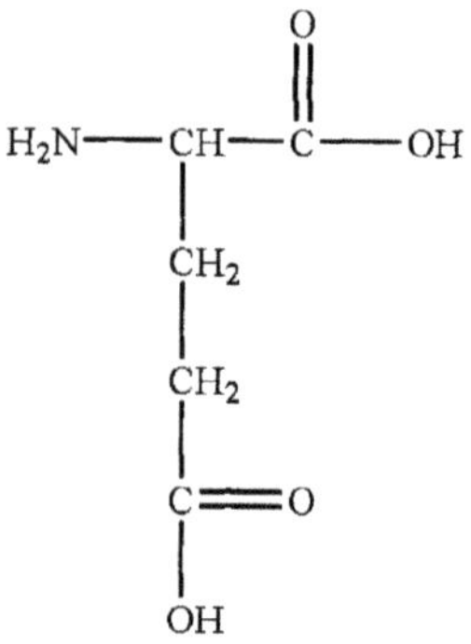

glutamic acid   glu   E

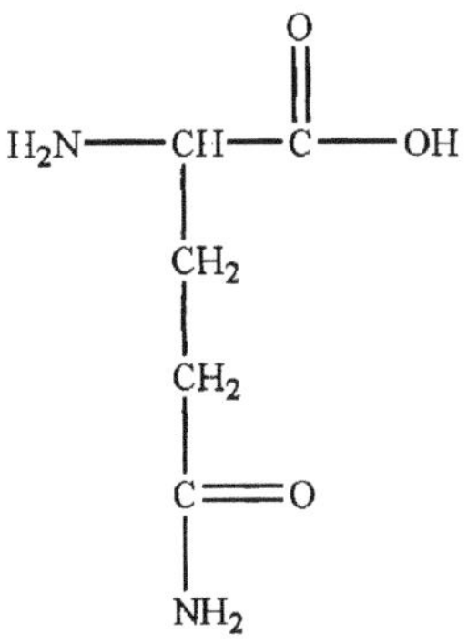

$$
\begin{array}{c}
\text{O} \\
\| \\
\text{H}_2\text{N} - \text{CH} - \text{C} - \text{OH} \\
| \\
\text{CH}_2 \\
| \\
\text{CH}_2 \\
| \\
\text{C} = \text{O} \\
| \\
\text{NH}_2
\end{array}
$$

glutamine   gln   Q

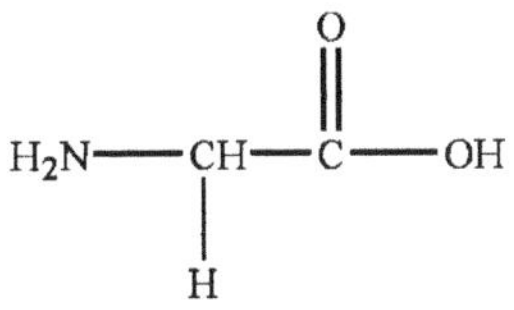

$$
\begin{array}{c}
\text{O} \\
\| \\
\text{H}_2\text{N} - \text{CH} - \text{C} - \text{OH} \\
| \\
\text{H}
\end{array}
$$

glycine   gly   G

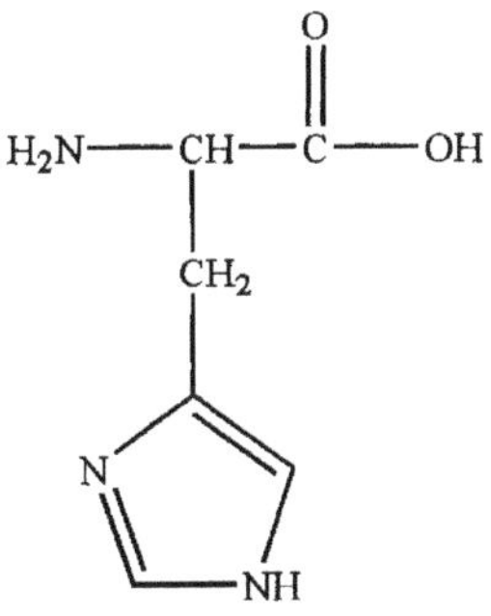

$$
\begin{array}{c}
\text{O} \\
\| \\
\text{H}_2\text{N} - \text{CH} - \text{C} - \text{OH} \\
| \\
\text{CH}_2 \\
\end{array}
$$

histidine   his   H

$$
\begin{array}{c}
\text{O} \\
\| \\
\text{H}_2\text{N} - \text{CH} - \text{C} - \text{OH} \\
| \\
\text{CH} - \text{CH}_3 \\
| \\
\text{CH}_2 \\
| \\
\text{CH}_3
\end{array}
$$

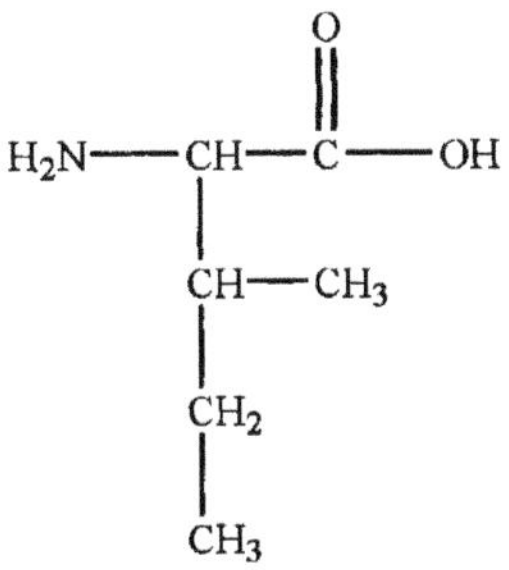

isoleucine   ile   I

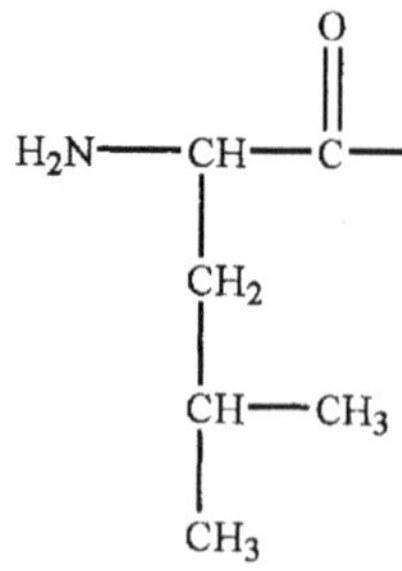

$$H_2N-CH-C(=O)-OH$$

with side chain:

$$CH_2$$
$$CH-CH_3$$
$$CH_3$$

leucine   leu   L

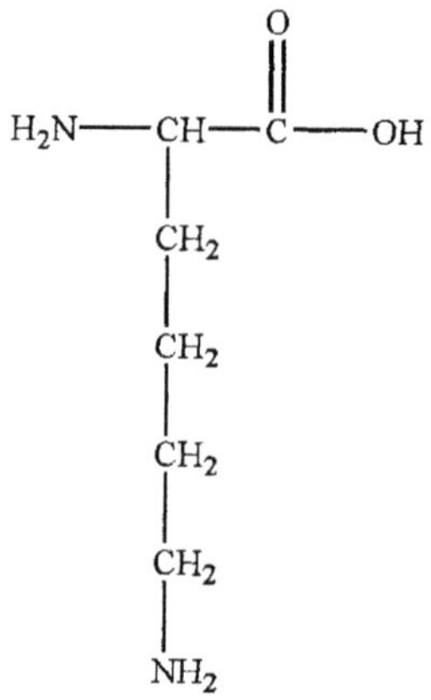

$$H_2N-CH-C(=O)-OH$$

with side chain:

$$CH_2$$
$$CH_2$$
$$CH_2$$
$$CH_2$$
$$NH_2$$

lysine   lys   K

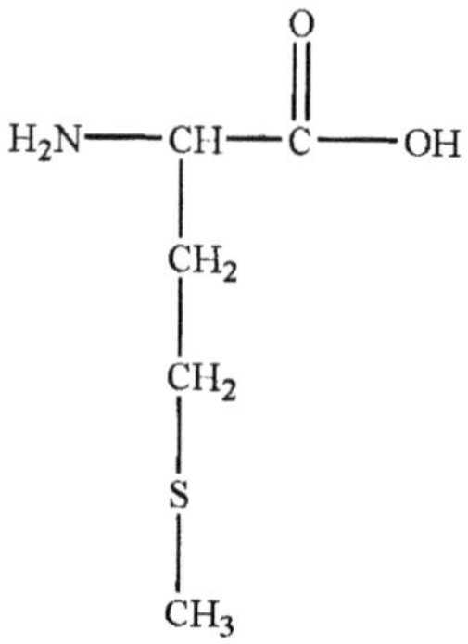

$$H_2N-CH-C(=O)-OH$$

with side chain:

$$CH_2$$
$$CH_2$$
$$S$$
$$CH_3$$

methionine   met   M

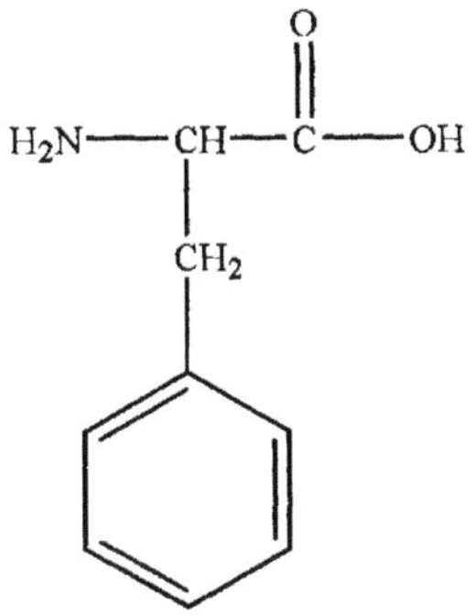

phenylalanine   phe   F

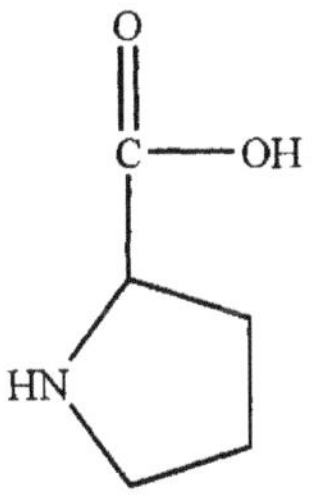

proline   pro   P

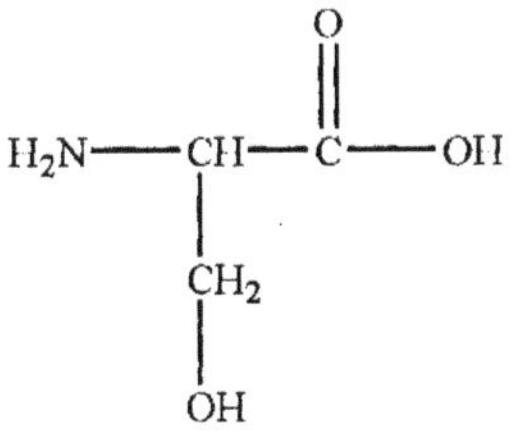

serine   ser   S

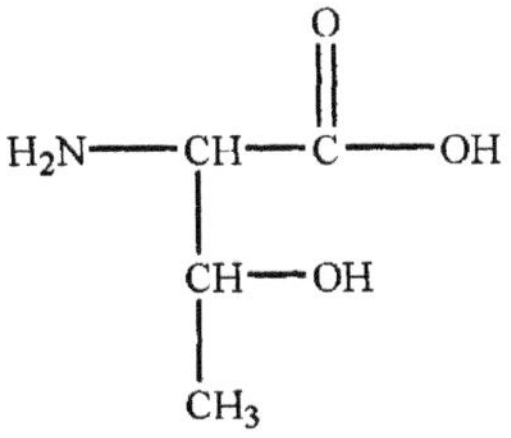

threonine   thr   T

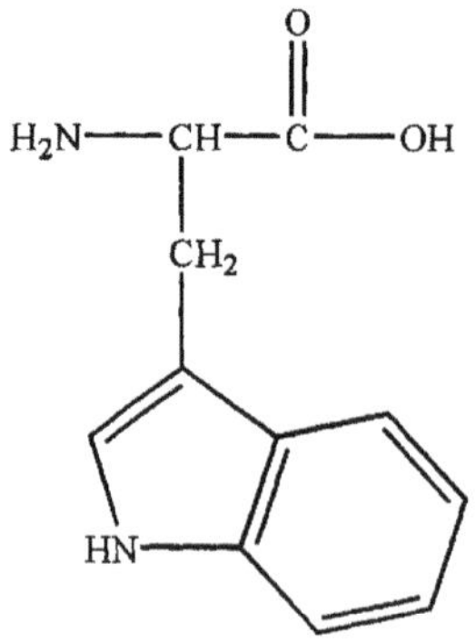

tryptophan	trp	W
tyrosine	tyr	Y
valine	val	V

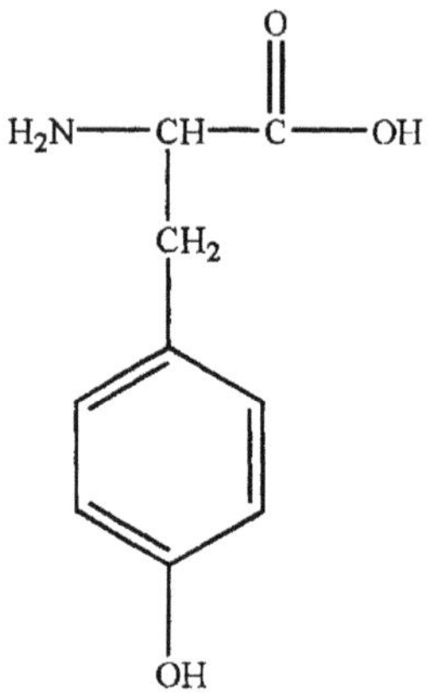

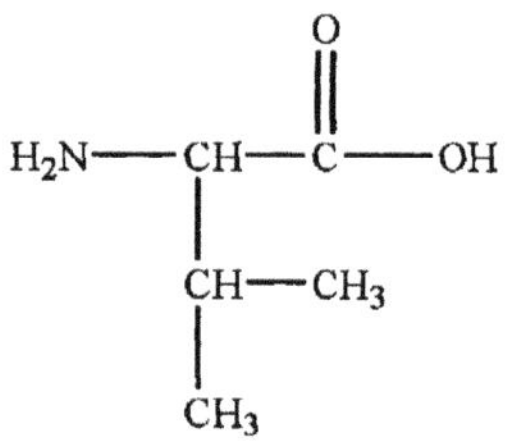

CHAPTER 1

The Ions

1.1 INTRODUCTION

Only sodium (Na), potassium (K), magnesium (Mg) and calcium (Ca) among the Group 1 and 2 elements are essential in biological systems. Some of the other s-block elements are used in medicine (*e.g.* lithium, Li and barium, Ba) and/or occur as minor (but useful) contaminants in calcium biominerals (*e.g.* strontium, Sr).

The metal ions with which we are involved in this book (Na^+, K^+, Mg^{2+} and Ca^{2+}) are less striking than many of the other metal ions that have significant biological roles. Metal ions such as iron and copper are coloured in solution, have more than one common oxidation state and are strongly coordinating, all properties lacking in the s-block metals. Nevertheless, the very lack of these properties enable them to be the workhorses in many biological processes. These metals display only one stable oxidation state resulting from the loss of one (M^+, Group 1) or two (M^{2+}, Group 2) s-electrons. This enables the metal ions to move around the cell without any danger of being oxidized or reduced. In this way they play many vital, complex and intriguing roles. They aid in neurologic and neuromuscular conduction, help regulate pH, maintain osmolality of body fluids, are involved in muscle contraction and are required for the functioning of many enzymes. They are components, calcium in particular, of many of the solid structures in most organisms.

The Heart Says It All

Animal tissues can be kept alive for experimentation for a short time by immersion in a buffered solution that mimics the ionic composition of animal plasma (Ringer's solution). A graphic illustration of the importance of these ions is shown by the composition of a cardioplegic solution used to preserve a donor heart (for 4–6 hours) prior to a transplant. The solution typically contains NaCl, 144 mM; KCl, 20 mM; $MgCl_2$, 16 mM; $CaCl_2$, 2.4 mM and procaine (a local anaesthetic that blocks nerve sodium channels, see 4.6), 1.0 mM at a pH of 5.5–7.5 at 4 °C.

1.2 OCCURRENCE

1.2.1 Earth's Crust

The s-group metals Na, K, Mg and Ca are, after Al and Fe, the most abundant metals in the earth's crust.

An excess of sodium salts in the soil can cause severe problems in crop production. This is because an influx of sodium ions into plant cells can upset critical biochemical processes by competing with potassium ions in membrane transport and enzyme activation, and can also cause osmotic stress. It has been estimated that about one-third of the world's irrigated land is unsuitable for growing crops because of high sodium ion contamination. Calcium ions, on the other hand, can be beneficial in that they help to maintain or enhance the selective absorption of K^+ ions by plants.

A Possible Solution to Salt Contamination of Soil

Shrubs of the genus *Plantago* produce seeds that are used as a food for birds and also as a bulk-forming laxative. Plants, even those of the same species, have differing abilities to tolerate salty soil. *Plantago maritima*, but not *P. media*, is salt tolerant. *P. maritima* (but not *P. media*) contains an Na^+/H^+ antiport protein that permits the interchange of Na^+ and H^+ ions (3.6.1). This allows the plant to sequester Na^+ ions in the large intracellular vacuole and away from the cytosol where salt interferes with metabolic processes. This represents one way of defeating a saline environment.

A salt-tolerant *Arabidopsis* has been engineered by overexpressing a single endogenous gene, AtNHX1, which encodes for an Na^+/H^+ antiport protein. Similarly, tomato plants have been genetically engineered to produce high levels of an Na^+/H^+ antiport. These plants flourish in 200 mM salt water, and produce good-looking and tasty fruit that is low in sodium even though the leaves accumulate large amounts of the ion. It could now be possible to produce a whole array of salt-tolerant crop plants, which would enable the use of seawater for irrigation.

1.2.2 The Seas

Most (about two-thirds) of the earth's surface is covered with water, and in the seas the s-group metal ions specified above are the major cations present. On average, there is 3.5 g of salts per 100 mL of seawater. This rises to 4% in the Mediterranean and off the southwest coast of Crete, where near-dry conditions 3–5 million years ago has produced a brine pool with a very high $MgCl_2$ concentration. The great salt lake in Utah is saturated with salts and NaCl crystallizes on the shore. In the Dead Sea, $CaSO_4$ crystallizes out. Nevertheless, certain creatures, tiny crustacean brine shrimp (*Artemia*) for example, but not fish, can survive in these conditions. Indeed, *Artemia* cannot live for long in fresh water. Coastal evaporation ponds are a commercial source of *Artemia*, which is used as fish food.

1.2.3 Biological Materials

The s-group metal ions Na, K, Mg and Ca are found in most cells in mM concentrations, see Table 1.1. It is often very difficult to measure the intracellular concentrations, particularly of the free ions. Inorganic and organic anions, which are lower in concentration outside the cell than inside, maintain cell neutrality.

One can draw one's own conclusions about the significance, if any, of the similarity between the concentrations of these ions in extracellular

Table 1.1 *Concentrations (mM) of extracellular and intracellular ions in squid axon and a representative mammalian cell*

	Intracellular		Extracellular		
Ion	*Squid*	*Mammalian*	*Squid*[a]	*Mammalian*	*Seawater*
Na^+	50	5–15	440	145	460
K^+	400	140	20	5	10
Mg^{2+}	10	1–3[d]	55	0.5–2.0[d]	54
Ca^{2+}	10^{-4}[a]	10^{-4}[b]	10	2.5–5.0	10
Cl^-	60	4	560	110	540

[a] Free Ca^{2+} ion, total Ca^{2+} ion = 0.4 mM [b] Free Ca^{2+} ion, total Ca^{2+} ion = 5 mM [c] A variety of marine animals have body fluid ion concentrations similar to those of the squid. [d] Free Mg^{2+} ion. Total Mg^{2+} ion is significantly higher.

media and in seawater in which life may have evolved. Although there are significant differences between ion concentrations in vertebrates and invertebrates, the *differences* between the solute composition of the cells and of the extracellular environment, with the glaring exception of Mg^{2+}, persist for squid and for mammals. These concentration differences are used to carry out many of the vital tasks that are necessary to maintain life.

1.3 COORDINATION CHEMISTRY

1.3.1 Ion Sizes

There is some disagreement as to the sizes of the ions of the s-block (particularly of Li^+). However, Pauling's values have, by and large, stood the test of time and are shown in Table 1.2.

We will find that ion size is particularly relevant when the transfer of ions through channels is being considered. For example, a long-standing vexing question arises as to how a dehydrated Na^+ ion, which is substantially smaller than a K^+ ion, can be rejected by the very selective K^+ channel. Earlier proposals have now been strongly supported and extended by an X-ray crystallographic examination of the K^+ channel (KcsA) cloned in mg amounts from the membrane of the bacterium *Streptomyces lividans*. For more about the channel structure and passage of K^+ through the pore see 3.4.2. The constriction (~ 3Å across) at the selectivity filter demands that all coordinated H_2Os are stripped from the hydrated ion for passage of the ion through the channel. An ideal fit and coordination by the channel O atoms compensate for the associated energy loss. These criteria apply to K^+ ions much better than to Na^+ ions. This accounts for an ~10^2 preference for K^+ ion entry (see Figure 1.1).

1.3.2 Donor Atoms and Strength of Binding

The s-block cations are highly electropositive and in the formation of complexes with smaller ligands have a preference for those with O-donor atoms. This generalization applies equally well to their coordination in metallobiomolecules where the ligands are polypeptides, proteins, nucleo-

Table 1.2 *Sizes of dehydrated s-block ions (radii in Å)*

Li^+	Na^+	K^+	Rb^+	Cs^+	Mg^{2+}	Ca^{2+}	Cl^-
0.60	0.95	1.33	1.48	1.69	0.65	0.99	1.81

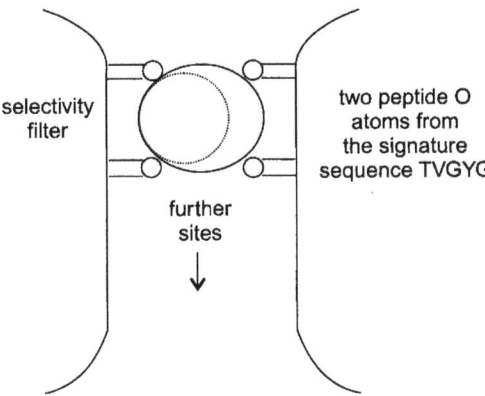

Figure 1.1 *One of the four potential metal ion sites in the selectivity filter of the KcsA channel. Two of the subunits are depicted. A third subunit is positioned in front of and the fourth is behind the drawing. The four subunits contribute a total of eight carbonyl O atoms from the peptide backbone that coordinate to the K^+ ion. These are arranged as a square antiprism around the metal ion, i.e. a twisted cube where the top is rotated 45° relative to the bottom. The protein structure forces the carbonyl oxygens into favourable positions for binding to K^+ but not Na^+ ions (dotted outline). In addition, the eight-fold geometry around the K^+ ion closely matches that of $K(H_2O)_8^+$ ion that enters the selectivity filter from a central cavity (3.4.2)*

tides (1) and polynucleotides. In these ligands the O atoms are provided by the RCO_2^- and OH groups from the side chains of amino acids, $-C=O$ from peptide bonds and PO_4^{3-}, HPO_4^{2-}, H_2O or OH^- moieties. Some illustrative formation constants and comments are shown in Table 1.3 (see Structure (1)).

(1)

Many enzymes require the presence of a +1 cation, particularly K^+ (2.2.2), for activity. The interactions of Na^+ and K^+ with these enzymes is unlikely to be strong and therefore the metal ion is not an integral part of

Table 1.3 *Log of formation constants at 25 °C and I = 0.1 M*

	AMP^{2-}	ADP^{3-}	ATP^{4-}	
Mg^{2+}	1.6	3.2	4.1	

These values indicate that γ, β and possibly α-phosphate groups in ATP^{4-} bind to Mg^{2+}.

	Na^+	K^+	Mg^{2+}	Ca^{2+}
ATP^{4-}	1.1	1.0	4.1	3.8
$EDTA^{4-}$	1.6	0.8	8.8	10.6

These values indicate that the strength of binding to a multidentate ligand is $Ca^{2+} \geqslant Mg^{2+} \gg Na^+, K^+$.

the enzyme. In the several crystal structures that have been solved, the first coordination sphere around Na^+ or K^+ has been occupied predominantly by oxygen atoms.

The atoms that coordinate to Mg^{2+} and Ca^{2+} are a much more important consideration as these ions form much stronger complexes, see Table 1.3. A startling illustration of the stronger binding power of Ca^{2+} over Na^+ is in the behaviour of mucus. Mucus is a slimy material rich in the glycoprotein mucin. It is an effective lubricant in the mouth, stomach, *etc.* and aids in the transfer of dirt and bacteria from the lungs to the mouth. It is stored as compact granules in the secretory vesicles within goblet shaped cells. The condensed polymer gel is maintained by high Ca^{2+} (and H^+) ion concentrations. These ions bind to the negative charges on the sulfate and sialic termini (acidic sugars found in many glycoproteins) of the long mucus strands. Mucus is released by intracellular calcium ($[Ca^{2+}]_i$) regulated exocytosis (5.1). The tightly packed mucin rapidly springs open as the Ca^{2+} ion concentration falls and is replaced by Na^+ ions. The sodium ions cannot hold the mucin together, so due to negative charge repulsions, mucin goes to the expanded state in an explosive release. The mucin network in the giant granules of the slug expands approximately 600-fold in 20–30 ms!

Suffocating Mucus is Used as a Weapon by Hagfish

The hagfish is an eel-like scavenger that can attack and suffocate predator fish using mucus. Hagfish have about 150 slime glands along their sides. When provoked the gland contents are ejected rapidly into the surrounding sea. The contents include the mucin-packed vesicles that break open. The ejected slime (about 5 g), which is a mixture of mucus (expanded mucin) and thin strong fibres that act as a scaffold, is an effective trap. Hagfish can strip fish (live, dead or dying!) so entangled using a rasp-like tongue.

Oxygen donor atoms are strongly preferred by Mg^{2+} ions and these emanate from phosphate groups in DNA, RNA, ribozymes, *etc.* Sometimes this coordination is reinforced by attachment of a nucleotide base N atom. This interaction is probably the most important impact made by Mg^{2+} ions in biology.

Diagnoses of Illnesses Could be Simplified

In the polymerase chain reaction (PCR) a target DNA segment with a genetic sequence specific for a particular organism can be amplified millions-fold using a specific primer (oligonucleotide), DNA polymerase and the four deoxyribonucleoside triphosphates.

If the blood of a patient contains the target DNA arising, for example, from a bacterium, then by applying PCR, conversion of the nucleotides into DNA will occur and the bacterial-induced illness can be diagnosed. A method suggested for the detection of the PCR-amplified material uses the release of Mg^{2+} ions (and the attendant conductivity increase) that accompanies the conversion of the nucleotides to DNA. The nucleotides, with three phosphate groups, bind more strongly to Mg^{2+} ion than does DNA, with only one phosphate group per base.

The binding of four N atoms from the corrin ring in chlorophyll to Mg^{2+} is a rare example of nitrogen coordination in a biomolecule with an s-block cation.

1.3.3 Calcium Binding Domains

Only O-donor atoms are known so far for Ca^{2+} ion binding to biomolecules. Two Ca^{2+} ion-binding motifs in particular are favoured in hundreds of proteins with widely varying functions. These are the EF hand, typified by the calcium site in calmodulin, and C_2-domains. The binding of the latter to phospholipids greatly enhances the strength of Ca^{2+} ion binding (see below).

There are two globe-shaped regions connected by a seven-turn α-helix at the N- and C-termini of calmodulin. Each globular region contains two Ca^{2+} ion binding sites. The amino acid sequence at each Ca^{2+} ion binding site is highly conserved in the proteins from various animals and plants.

Each calcium ion is seven-coordinated, with the O atoms provided by side chain carboxylate (glu, asp), carbamoyl (gln, asn) and alcohol or peptide $C=O$ (thr, ser, tyr) groups and H_2O. These originate from the 12 residue-containing loop in the E helix–loop–F helix motif (also called the E–F hand), that is also found in other calcium binding proteins (2.5). The E helix–loop–F helix motif contains a sequence of amino acids that is a useful signature for locating similar new proteins. This was one of the first structural motifs to be recognized in protein structures.

The other important Ca^{2+} binding entity is a so-called C_2-domain. This domain also binds to phospholipid and this combination enhances Ca^{2+} ion binding by 10^3-fold. In turn, Ca^{2+} ions promote membrane binding by C_2-domains. The C_2-domain, about 130 residues in length, is present in a wide range of proteins, *e.g.* protein kinase C, cytosolic phospholipase A_2, PLCδ1 (3.5.2) and synaptotagmin (2.5.3). In the C_2-domain, multiple Ca^{2+} ions bind in clusters. The residues involved are often aspartate side chains acting as bidentate ligands. PLCδ1 contains both E-F-hand and C_2 domains as well as catalytic sites.

1.3.4 Geometry of Binding

The regular geometries encountered with small complexes of these metals (octahedral coordination is most favoured) are not generally duplicated in metalloproteins because of the constraints imposed by the protein structure. Sometimes the s-block metal ion may even be compelled to bind to an N-donor, which has been forced into a coordination position by the protein.

Dialkylglycine decarboxylase (DGD) is a pyridoxyl phosphate-dependent enzyme that catalyses decarboxylation and transamination in a catalytic cycle (Equation 1).

$$(CH_3)_2C(NH_2)COOH + CH_3COCOOH \rightarrow$$
$$CH_3CH(NH_2)COOH + CO_2 + (CH_3)_2CO \quad (1)$$

Crystal structures of the enzyme with either a K^+ or an Na^+ ion in position near the active site have been very informative about the geometries and the roles of the univalent metal ions in the catalysis. In DGD, six oxygen atoms are octahedrally disposed around the K^+ ion (M–ligand ~ 2.73 Å) whereas only five oxygens form a distorted trigonal bipyramid around the Na^+ ion (M–ligand ~ 2.33 Å), Figure 1.2.

The change of geometry around the Na^+ ion induces marked changes in

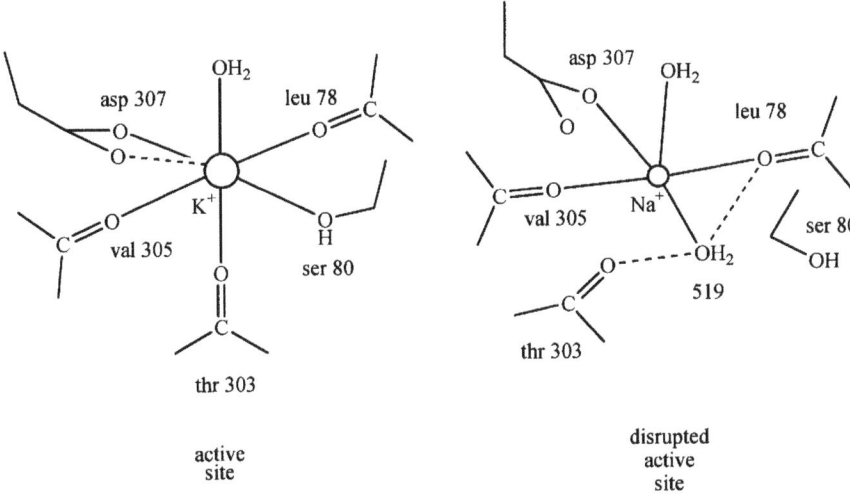

Figure 1.2 *The coordination of K^+ and Na^+ ions near the active site of DGD. The coordination sites are formed by two loop segments that are separated by approximately 220 residues. The smaller Na^+ ion and shorter $Na^+ - O$ bond lengths (compared to those of the K^+ ion) require a reduction in coordination number from six to five. This leads to the replacement of coordinated thr303 and ser80 in the K^+ form, by one H_2O in the Na^+ form. This disrupts the orientation of ser80 and in turn those of tyr301 and gln52 (not shown), which are peripheral to, but part of the extensive active site. The Na^+ form of the enzyme is thereby inactive*

the orientation of tyr-301, which probably interacts with bound substrates, and in ser-80, which is a K^+ but not an Na^+ ligand. These changes disrupt the nearby active site and can be used to rationalize the facts that Na^+ ion is an inhibitor and K^+ ion an activator of the enzyme.

The ionic radius of Mg^{2+} is much smaller than that of Ca^{2+} (Table 1.2). Usually the protein is not flexible enough to adjust the coordinating O atoms to the smaller Mg^{2+} ion. As a consequence, fewer O atoms from the protein are likely to coordinate to the Mg^{2+} ion (preferred coordination number of six) without strain, compared to the Ca^{2+} ion (coordination numbers of 6–8 common). This is the reason that the Ca^{2+} ion forms the stronger, more diverse complexes. The differences in the geometries exhibited by Ca^{2+} compared to those of the Mg^{2+} ion may cause different structural perturbations when each of these ions is associated with an enzyme. This idea has been used to rationalize the 10^{-3} difference in rates when an Mg^{2+} ion is replaced by Ca^{2+} ion in the reaction catalysed by isocitrate dehydrogenase in which Mg^{2+} is the endogenous ion *in vivo* (Equation 2).

$$\text{isocitrate} + \text{NADP} \longrightarrow \text{oxalosuccinate} \longrightarrow$$

$$\alpha\text{-ketoglutarate} + \text{NADPH} + CO_2 \quad (2)$$

1.3.5 Kinetic Lability

The hydrated Na^+, K^+ and Ca^{2+} cations are extremely labile – that is their coordinated water molecules exchange with solvent water at rates in excess of $10^8\ s^{-1}$. The replacement of coordinated water by ligands is therefore facile and unlikely to be a rate-limiting step. This is particularly important if the triggering action of Ca^{2+} ions is to be effective. In contrast the water exchange rates for Mg^{2+} ions are much slower ($\sim 10^5\ s^{-1}$).

1.4 SOLUBILITY OF SALTS AND ATTENDANT PROBLEMS

Most common sodium and potassium salts are soluble in water, although the solubilities cover a wide range. Therefore the presence of Na^+ or K^+ ions even in high concentrations in the cellular milieu, which contains several anions, usually causes no precipitation problems. An exception to this is in *hyperuricemia* where the offending anion is urate (2) (2,6,8-trioxypurine anion).

A Pain in the Big Toe

In higher primates the final product of purine degradation is uric acid, which is excreted in the urine. The overproduction, or impaired elimination, of uric acid leads to elevated amounts in serum (hyperuricemia) and urine. This results in the deposition of sodium urate crystals in extreme joints, *e.g.* the big toe, which is the inherited disease gout. A mixture of uric acid and sodium urate (kidney stones) can promote kidney disease. These problems only occur in humans, apes and the Dalmatian dog (!) because they lack or are unable to use the enzyme uricase, which in other mammals catalyses the further degradation of uric acid to very soluble allantoin (3).

(2)   (3)

The compounds of Mg^{2+} and Ca^{2+} ions with mononegative anions, *e.g.* chloride, are usually water soluble, but those with multicharged anions, *e.g.* phosphate, are often insoluble. Whereas this insolubility is essential with certain biomolecules (consider bone!), it raises serious problems in other cases. The precipitation of calcium phosphate in the cytosolic space is prevented by the maintenance of a low intracellular concentration ($<\mu M$) of free Ca^{2+} ion. This is accomplished by using a complex network of Ca^{2+} ion binding proteins, transporters and channels (3.8). The insolubility of calcium oxalate is the main reason for oxalate poisoning. The oxalate anion is very toxic since it can limit the bioavailability of Ca^{2+} ion by precipitating it as calcium oxalate. If the Ca^{2+} ion concentration, for example in the blood of a mammal, falls to one half its normal value the result is severe or fatal tetanic cramps. Large amounts of spinach consumed daily can lead to the precipitation of calcium oxalate and calcium deficiency, while eating an excess of rhubarb leaves, which contain even more oxalic acid, can lead to death (hypocalcemia). The oxalate anion was found in old hat paints and was used as the murder weapon in Agatha Christie's *Easy to Kill*.

Disastrous Cattle Drive

The plant halogeton, common in the western United States, contains up to 30% dry weight of oxalate. It is avoided by grazing cattle except in times of acute food shortage. During and following one trail drive, 16 of 680 range cattle suffered acute halogeton poisoning (which results in coma and death). Crystals of calcium oxalate were found in the renal tubes of the dead animals. In humans about 75% of urinary stones contain mainly calcium oxalate so that the medical implications are enormous.

Severe burns and scars that are slow to heal are caused by contact of hydrofluoric acid with the skin. Hypocalcemia (Table 2.3) also results and these effects arise from the precipitation of calcium fluoride. Copious washing with water and regular applications of calcium chloride or gluconate solutions are effective treatments.

1.5 ASSAY

1.5.1 Solution

To more easily assess their roles, it is vital to be able to accurately determine the concentrations and changes in concentrations of ions in a variety of environments. The determination of the *extracellular* concentrations of the s-block metal ions in biological fluids presents few difficulties. Standard methods in use include atomic absorption, flame photometry and ion-selective electrodes. The determination of *intracellular* ion concentrations, and of small changes in these concentrations, present big problems, however. Ion-sensitive microelectrodes have been used to measure the intracellular concentrations of Na^+, K^+, Mg^{2+}, Ca^{2+}, H^+ and Cl^- ions (even as low as 10 nM), but the method is difficult to use, has a much slower response time than spectral methods and is generally invasive (cells are impaled), which can lead to ion movement. Non-invasive ion probes placed near the plasma membrane for measuring weak extracellular voltages associated with ion flux have recently been described (6.5.2).

1.5.2 Determination of Intracellular Calcium Ion Concentrations

Measuring intracellular ion concentrations using fluorescence spectroscopy is now the usual method of choice and has been used for all the ions. It's value and potential will be illustrated for the determination of $[Ca^{2+}]_i$. We will see that small changes in cytosolic $[Ca^{2+}]$, *i.e.* on the μM level, are used as intracellular signals in many biological processes. The detection of these small concentration changes has been a vexing problem since the qualitative roles of calcium ions as ubiquitous and critical messengers were established. Furthermore, the assay for very small $[Ca^{2+}]$ must be accomplished in a sea of mM Na^+, K^+ and Mg^{2+} ions (Table 1.1). Fluorescent dyes are the chemical probes used most often for the detection of Ca^{2+} ions. Fluorescence is very sensitive, is specific, has a rapid time resolution and is amenable to microscopic detection. However, any upset of cellular physiology by the dye must always be assessed.

In 1967, the first reliable determination of $[Ca^{2+}]_i$ was made *via* fluorescence spectroscopy using the photoprotein aequorin, which was injected into the giant muscle fibre of the barnacle (see Figure 6.7). Since the 1980s the use of this photoprotein has been largely replaced (see jellyfish note however) by synthetic dyes whose fluorescence characteristics, like those of aequorin, are greatly changed when they bind to Ca^{2+} ions, even at very low concentrations and in the presence of other ions. Their use for the determination of the concentrations of the other ions has so far been less applied.

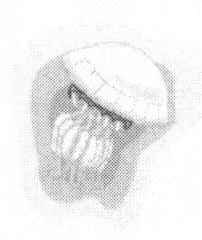

Jellyfish Protein Sheds Light on Calcium Ion Movement

The blue-green glow of the jellyfish *Aequorea victoria* and *A. forskalea* that live in the Puget Sound is used both to frighten off predators and to attract prey. The bioluminescence from the margin of the jellyfish's umbrella arises from:

- A blue luminescent emission when the protein aequorin (which contains four E-F hands (1.3.3)) binds to Ca^{2+} ions (the mechanism of binding is still uncertain).
- A green fluorescence re-emission when a green fluorescence protein (GFP) absorbs the blue light from aequorin. A luminophore, coelenterazine, and O_2 are also required.

The gene encoding for the photoprotein has been isolated and cloned and incorporated into specific organelles (*e.g.* ER, SR), where it can be used to monitor Ca^{2+} ion concentrations in the targeted organelle. Using recombinant aequorin reconstituted in thin live *E. coli* cells, the $[Ca^{2+}]_i$ is regulated at a concentration of about 100 nM, a value similar to that in resting eukaryotic cells. Although outside the province of this book, it is worth noting that proteins can be fluorescently labelled by fusing their genes with that for GFP and expressing the resulting hybrid.

The use of fluorescent dyes is now the most widely employed method for the measurement of $[Ca^{2+}]_i$. In many of these dyes there is an EGTA backbone, which provides the strong Ca^{2+} ion chelation centre, condensed with a fluorescent moiety. Fura-2 and indo-1, (4 and 5), are two of the most popular dyes.

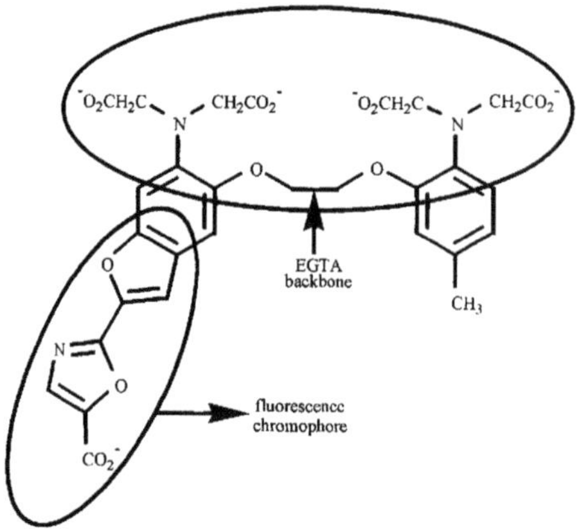

(4) 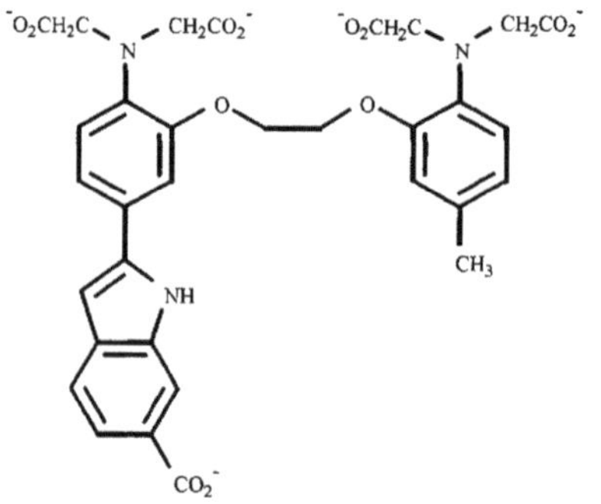   (5)

1.5.3 Procedures

In the presence of Ca^{2+} ions the fluorescence of fura-2 and indo-1 is enhanced and the excitation wavelength is shifted to lower wavelengths. The ratio of the amplitudes at two different wavelengths gives the value of the Ca^{2+} ion concentration independent of dye concentration, tissue thickness and absolute intensity of the excitation source. The ester derivatives are used since they can pass through the membrane. Hydrolysis of the ester groups to form the dyes is catalysed by esterases in the cell. The dyes are injected into the sample (which can be living tissue) and placed on the stage of a fluorescence microscope (similar to a conventional microscope, but using excitation wavelengths that induce fluorescence in the sample). A number of methods of increasing sophistication (and expense!) are available for imaging individual planes of focus in even quite thick tissue samples. In this way a few thousand molecules can be examined in cubic micrometers. A recent development is to encapsulate the fluorescent indicator in a polyacrylamide matrix thus forming spherical nanosensors 20–200 nm in diameter. The matrix allows ions to diffuse easily and bind to indicator molecules, but does not allow indicator molecule release, which protects cellular contents from possible undesirable interactions with dye. Response times are rapid (<1 ms) and the nanosensor can be picoinjected into single cells *in vivo* and pH and $[Ca^{2+}]$ changes can be measured.

Examples of the use of fluorescent dyes include observations of the following: Ca^{2+} ion-induced exocytosis (Figure 5.2), the rise and fall of Ca^{2+} ion concentration in the muscle cells during each heartbeat of a newborn rat and the transient increase of $[Ca^{2+}]$ in sea urchin cytoplasm

15–36 seconds after the injection of sperm (2.5.3). The use of these indicators has had a tremendous impact on our understanding of the details of Ca^{2+} ion movement within cells and between cells and extracellular fluids. One of the most startling recent findings has been the observation of intracellular heterogeneity arising from calcium ion 'waves' and 'sparks'.

Using a mixture of dyes allows the analysis of the concentrations of different entities simultaneously and thus their interrelationships. For example, inositol-1,4,5-triphosphate (IP_3) mobilizes Ca^{2+} ion from storage using an IP_3 receptor (Figure 3.6). This results in the activation of Ca^{2+}-dependent cellular events such as muscle contraction, secretion, *etc.* (6.2.1). By monitoring IP_3 (by using a probe tagged with GFP) and Ca^{2+} (using fura-2) at the same time in a single cell, it can be shown that the mobility of calcium ions (expressed as oscillating waves) occurs synchronous with changes in IP_3 concentrations. These approaches to concentration measurements herald the era of real-time cellular physiology observations.

1.5.4 Caged Calcium

A clever technique that allows the rapid generation of Ca^{2+} ions *in situ* can be used to activate Ca^{2+}-dependent physiological processes.

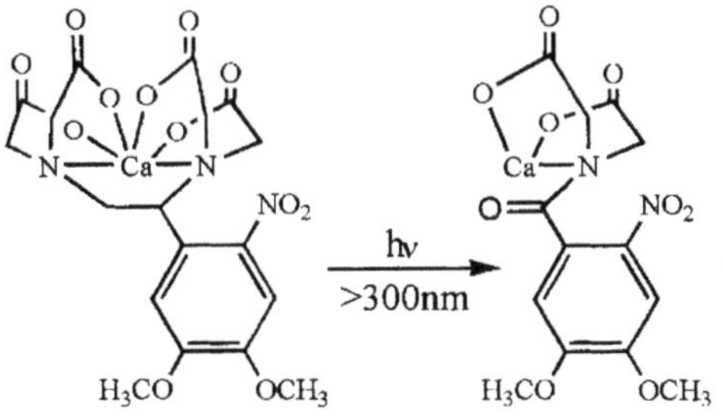

$$+ NH(CH_2CO_2^-)_2 + \text{free } Ca^{2+} \qquad (3)$$

Ida

The ligand used in the method, DM-n, binds strongly to Ca^{2+} ions *via* an EDTA moiety. Laser photolysis of this complex ($K_{assoc} = 2 \times 10^8 \text{ M}^{-1}$) containing a 'caged' calcium ion (6) partially breaks down the bound ligand to fragments that form weaker complexes (7) and Ida ($K_{assoc} = 3 \times 10^2 \text{ M}^{-1}$). Calcium ions are thus released within a millisecond (Equation 3). The Ca^{2+} ions generated can be used, for example, to stimulate the release of transmitter at a synapse, and the Ca^{2+} ion distribution in the system can be followed with an indicator (5.1).

Alternatively, the rise in tension in isolated toadfish swimbladder fibres (6.3.2) can be measured after rapid activation by Ca^{2+} ion generated by laser photolysis of a Ca-nitrophenylEGTA complex (8).

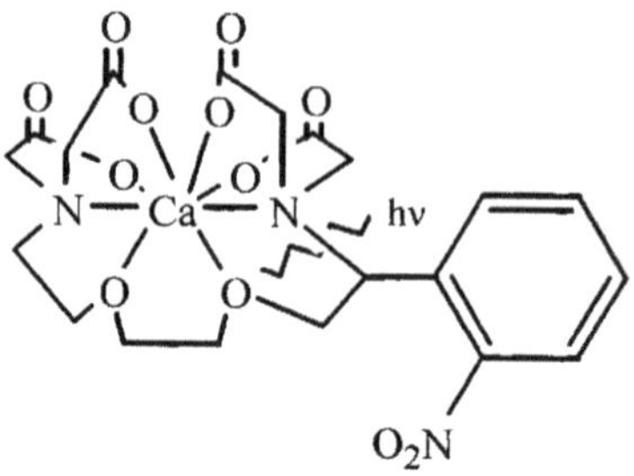

(8)

1.5.5 Solid State

The chemical composition of the calcite shells of the unicellular marine organism foraminifera has been used as a palaeothermometer to probe past ocean temperatures (8.1). Also, the make-up of the aragonitic otoliths (ear stones) of fish reflect the characteristics of the water in which the creatures live and can shed light on their migratory habits (8.4.1).

In both cases, a substantially pure sample of $CaCO_3$ must be analysed for trace amounts of impurities ranging from (per Ca^{2+} ion) 10^{-3} mol mol^{-1} (Mg and Sr) to as low as 10^{-7} (Ba and Cd). Accurate analytical methods must obviously be used. Until recently, elemental ratios in samples were determined indirectly by combining different analytical methods, usually atomic absorption or emission spectrometry. The introduction of methods for the simultaneous determination of several element/calcium ratios in small samples has been an important development. The combination of inductively coupled plasma with mass spectrometry or optical emission spectrometry uses samples as small as 25 μg with an observed accuracy of <0.5% for Mg/Ca or Sr/Ca and of <1% for Ba/Ca.

1.6 SUMMARY

Of the s-group elements, only sodium, potassium, magnesium and calcium are biologically significant. The properties that play important roles in their ability to function were detailed. These include the sizes of the ions and their coordination chemistry. Their concentrations within and outside the cell are important in cell behaviour and signalling. In recent years the ability to monitor intracellular metal ion concentrations, particularly that of calcium, has been important in assessing the functions of the ions in many biological events and especially in real-time changes.

CHAPTER 2

Biological Roles

In this section we cover the background for the main biological phenomena in which the metal ions are involved. It is not intended to be all encompassing, but rather to give a flavour of the tasks that the metal ions are able to carry out. The relatively large space given to the coverage of calcium attests to the dominance of, and more varied roles for, calcium in cellular biology. Indeed there is a journal devoted solely to cellular aspects of the element (*Cell Calcium*). In a number of instances the effectiveness of calcium cannot be duplicated by any of the three other metal ions. Nevertheless the number of publications from the mid-1960s to the present in 4000 biomedical journals (source, *NIH Medline*) relating to sodium is approximately the same as for calcium (over 250 000). They far exceed those for the other ions with which we are concerned, $K^+ >> Mg^{2+} >> Cl^-$, although this reflects the multiplicity of roles for calcium and sodium ions rather than their necessity.

2.1 SODIUM

2.1.1 Impulse Transmission

Sodium ions play a major role in impulse transmission by using the large difference in their concentration inside and outside the cell. Na^+ ions are responsible for the upstroke of the action potential in certain cells (4.4). All creatures attempt to acquire the requisite amount of Na^+ ions. Wars have been fought over the possession of salt deposits. The quality of the taste of salt attracts animals to salt licks and because many other ions are often found in $NaCl$ deposits, any mineral imbalances can also be corrected.

Most plants accumulate more K^+ ions than Na^+ ions. Herbivores therefore need to acquire Na^+ by other means. Porcupines, for example, annoy property owners by gnawing on wooden material. They do this to obtain Na$^+$ ions from NaNO$_3$, which is used as a curing compound in plywood. Moths and butterflies go to even more extreme lengths to acquire Na$^+$ ions.

Moths and Butterflies Guzzle Salty Water

Moths and butterflies (mainly male) procure sodium by 'puddling'. In this process they take up large volumes (twice their weight per minute) of the saltiest water they can find, *e.g.* at the edges of puddles. They then eject the water in spurts, typically through 20 anal jets per minute and persisting for as long as hours, to some distance away. The sodium ions are absorbed in the hindgut. There is a close correspondence between gain of Na$^+$ ions and loss of K$^+$ ions, presumably through the use of ion channels (e.g. Na$^+$/K$^+$ ATPase), but no change in [Mg^{2+}] or [Ca^{2+}] is observed. The males transfer Na$^+$ ions to the females during mating for incorporation into their eggs.

2.1.2 Solute Transport

The Na$^+$ ion is a major player in the control of cellular pH (3.7.1). The transport of most solutes across the membranes of the epithelial cell uses an Na$^+$ ion acting as a cotransporter or countertransporter, Figure 2.1.

Nutrient material (X) in the gut lumen has to pass through the cytosol of the epithelial cell to enter the bloodstream. A Na$^+$/X symporter (3.6.2) in the apical membrane and an X transporter in the basolateral membrane effect this transfer. Glucose (X) is moved by this process and Na$^+$ ions are also moved in the same direction *via* an amiloride-sensitive Na$^+$ channel (3.6.1) and Na$^+$/K$^+$ ATPase. The resultant Na$^+$ and glucose movement across the epithelium leads to the transfer of water in the same direction by osmosis. This occurs by diffusion across H$_2$O channels and even across a tight junction. This is the basis of the NaCl/glucose drip (Figure 2.1) that allows the effective transfer of water to dehydrated patients, *e.g.* those suffering from severe diarrhoea. Adding water to the lumen is nowhere near as effective or as rapid as the drip.

The amiloride-sensitive sodium channel is present in a number of epithelial cells. The channel has major roles in the control of blood pressure and blood volume, and has been implicated in cystic fibrosis.

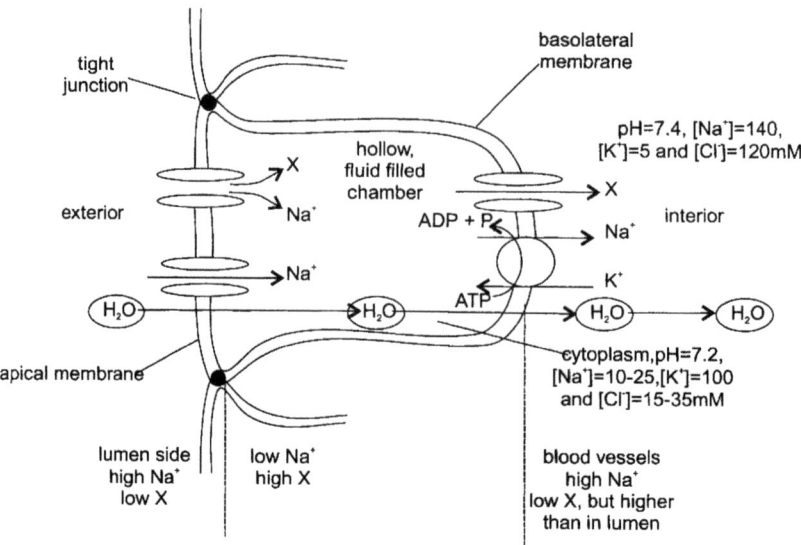

Figure 2.1 *Intestinal epithelial cell. Epithelial cells line hollow organs and are responsible for the exchange of material between the body and the exterior or body fluid compartments. They also maintain and regulate extracellular fluid volume and composition in both vertebrates and invertebrates. Generally, ions and neutral molecules are transported by a transcellular pathway through the apical membrane, cytoplasm and basolateral membrane. Different pumps and transporters are used in the two membranes, which are distinct and are separated by tight junctions that rarely allow movement of solutes between adjacent cells (paracellular pathway). The apical and basolateral portions of the cell membrane have different characteristics, the basolateral membrane favouring K^+ ion conduction and being the sole depository of the $Na^+/K^+ATPase$ channel. Recent evidence points to Mg^{2+} (but not Na^+, K^+ or Ca^{2+}) ion being resorbed through paracellular channels in kidney tubules driven by an electrochemical gradient across the tubule epithelium. Mutations in paracellin-1, which regulates this resorption, may be implicated in the wasting syndrome hypomagnesemia (Table 2.3)*

A Child's Salty Forehead Could Spell Danger

Cystic fibrosis is a serious human disease affecting about 1 in 2000 newborn babies in the United States annually. It affects primarily the lungs and exocrine glands. The principal symptom is a thick secretion of mucus that obstructs the airways, harbours bacteria and leads to chronic lung infections and general respiratory difficulties. A simple test for the condition is a high salt concentration in the sweat of the victim. The source of the problem is the gene that encodes for the cystic fibrosis transmembrane conductance regulator (CFTR). This is an epithelial voltage-independent chloride ion channel, which appears in many organs and regulates water and salt balance in the lungs and intestines. CFTR also acts as a regulator of an epithelial sodium ion channel (BR5A). Cystic fibrosis of varying severity arise from mutations in the gene, the most common and damaging of which is the deletion of a phenylalanine (ΔF 508) in a nucleotide binding domain (NBD). How the defect in ion transport triggers the lung problems of CF is still uncertain. Natural antibodies in cells lining the inner surface of the lungs, *e.g.* defensins, cannot combat infections in the presence of high NaCl concentrations.

An interesting (and so far unanswered) question is therefore: do sea creatures produce natural antibodies that are insensitive to salt? In addition, excessive activation of the CFTR in the gut by bacterial toxins can cause secretory diarrhoea. This is the second largest cause of infant mortality in the developing world.

2.2 POTASSIUM

2.2.1 Impulse Transmission

The potassium ion is a major player in the maintenance of membrane potential and is responsible for the downstroke of the action potential and for the return to resting potential (4.4).

2.2.2 Enzyme Activation

The complexing of metal ions by biomolecules is the basis of the many roles played by them. The metal–ligand interactions range from weak to very strong (Table 1.3). Monovalent cations are loosely associated with many biomolecules. It is through this weak binding that metal ions are able to activate a large number of enzyme catalysed reactions. This was first demonstrated in 1942 for the K^+ ion activation of pyruvate kinase. Since then the K^+ ion has been shown to usually be the most effective ion in circumstances where a monovalent cation is required for enzyme activity. Only with some of the extracellular enzymes is the sodium ion more effective. Most of the intracellular K^+ ions (and Mg^{2+} ions) are bound to ribosomes (small particles of RNA and protein, the sites of protein synthesis). This binding of K^+ and Mg^{2+} ions to nucleic acids reduces the electrostatic repulsion between the phosphate groups and thus stabilizes the structure.

2.3 MAGNESIUM

2.3.1 Impulse Inhibition

The concentration differences observed for Na^+, K^+ and Ca^{2+} ions inside and outside cells are not observed for the magnesium ion, which therefore has no role in nerve impulse transmission. It can, however, inhibit the process.

Anaesthetic Effect of Magnesium Ion

A rise in serum $[Mg^{2+}]$ can block the effects of Ca^{2+} ions and thus cause inhibition of acetylcholine release from motor nerve terminals. This has an anaesthetic effect and may underlie the observation that fast moving marine animals (*e.g.* Norwegian lobster) have a lower body fluid content of Mg^{2+} (9.3 mM) compared with that of a slow moving Maia crab, for example (44.1 mM). Elevated serum $[Mg^{2+}]$ is also a characteristic of hibernating squirrels, bats and hedgehogs where there is an increase of nearly 100% over control $[Mg^{2+}]$. The reason for this is still unclear, however.

It has been suggested that an anaesthetic effect of the ion reduces the painful sensations sometimes associated with the use of Mg^{2+} salts in the treatment of constipation and heartburn.

2.3.2 Interactions with Biomolecules

The Mg^{2+} ion has a much stronger affinity than monovalent cations for carboxylate and phosphoryl moieties and this plays a key role in cell wall structure, enzyme activation and the stabilization of DNA, ribosomes and related species.

2.3.3 Cell Wall Structure

The Mg^{2+} ion is used (with Ca^{2+} ions) to cross-link carboxylated and phosphorylated polymers in cell walls. This is the reason that ligands like EDTA can disrupt cell walls and cell–cell structures by chelating and ripping out the metal ion. The binding and bridging ability of Mg^{2+} and Ca^{2+} ions in biopolymers also accounts for the fate of marine colloids. These comprise about 30–50% of the dissolved organic material (which is largely anionic in nature) in seawater. Under the bridging influence of these metal ions contained in the marine colloids, the latter coalesce to form a large conglomerate that sinks.

2.3.4 Enzyme Activation

This is undoubtedly the *tour de force* of Mg^{2+} ions. It has been estimated that over 300 enzymes of widely varying types are activated by the Mg^{2+} ion. This takes place either through association of the Mg^{2+} ion with the substrate (particularly nucleotides) or by Mg^{2+} ion binding to the enzyme promoting allosteric activation. Often the two effects occur together, *e.g.* with reactions catalysed by adenylate cyclase, Equation 1, guanylate cyclase, Equation 2 and ATPases to name a few.

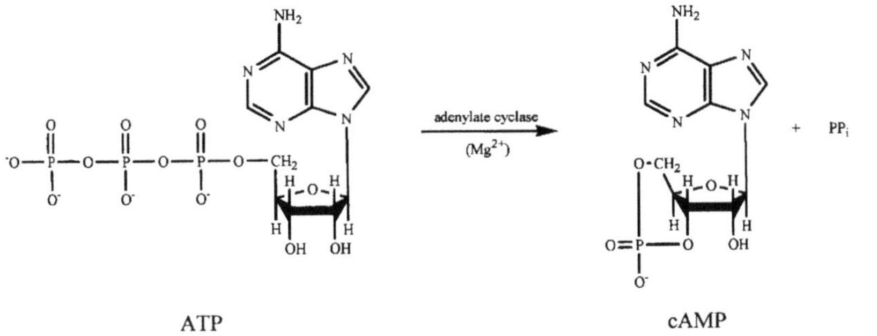

$$\text{ATP} \xrightarrow[\text{(Mg}^{2+})]{\text{adenylate cyclase}} \text{cAMP} \quad + \quad PP_i \tag{1}$$

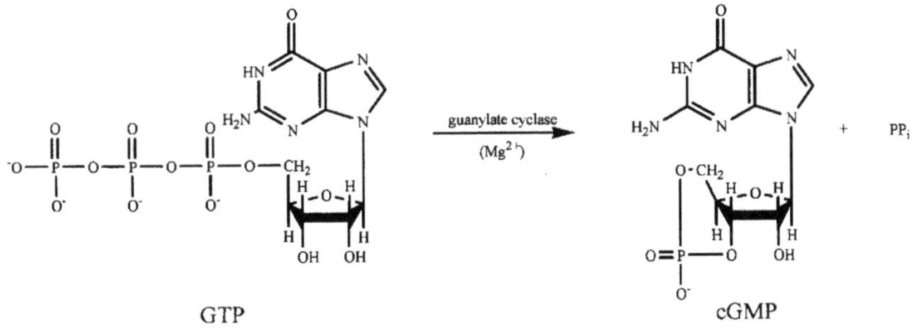

$$\text{GTP} \xrightarrow[\text{(Mg}^{2+})]{\text{guanylate cyclase}} \text{cGMP} \quad + \quad PP_i \tag{2}$$

The hydrolysis of ATP drives many biological processes. The binding of Mg^{2+} ion to ATP and to ATPases is essential for their successful operation. This contribution of the Mg^{2+} ion is sometimes overlooked in comparison with its more glamorous relative, the Ca^{2+} ion in, for example, muscle contraction (6.2.2). All enzyme catalysed reactions that involve phosphoryl transfer require an Mg^{2+} ion. One of the earliest enzyme activities must have been that of polynucleotide polymerases because the ability to accurately replicate the genome was necessary for the development of evolution. The magnesium ions must have been involved at the outset. In precursor RNA it is necessary to excise the introns from the exons that contain the coding sequences. The exons must then be accurately joined to make a functional messenger RNA. The magnesium ion is a critical component of the active site of the 'splicing' enzyme.

Some insight into the mode of action of the metal ion comes, for example, from consideration of catalysis by an RNA enzyme (*ribozyme*) called *Tetrahymena* Group I ribozyme. This enzyme recognizes a specific RNA sequence, cleaves it at two specific locations and then splices the two ends back together, thus removing the small cleaved fragment. It is difficult to diagnose specific roles of bivalent metal ions in such catalyses because a flood of metal ions do a number of things. They: (a) attach to negative charges of the RNA backbone, (b) help RNA to fold correctly and, (c) aid in catalysis. The important oxygen atoms in the binding (active) site were identified by replacing O atoms by S atoms. If the replaced oxygen atom is bound to a catalytically important Mg^{2+} ion then that modification will suppress activity since the Mg^{2+} ion coordinates more strongly to O atoms than to S atoms. By this means their role in (c) was deciphered. Two metal ions are involved and have important and different roles in catalysis (two metal ion mechanism). In our example, one metal ion interacts with an O atom on the leaving group and the other with the incoming nucleophile. Both also interact with one of the nonbridging O atoms on the transferred phosphoryl group, Figure 2.2.

Another enzyme catalysed reaction in which an analogous two M^{2+} ion catalytic domain is invoked is the regulation of cAMP and cGMP by

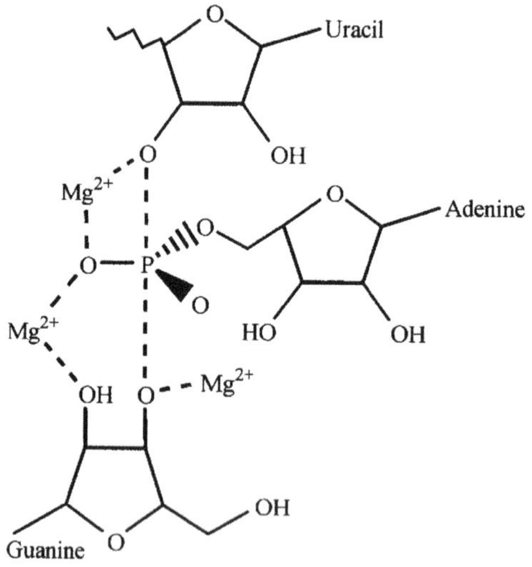

Figure 2.2 *The active site in the self splicing reaction catalysed by Tetrahymena group I ribozyme. Three Mg^{2+} ions are involved in the transfer of a phosphoryl group from the top nucleotide to the bottom nucleotide*

phosphodiesterases. As we shall learn, cAMP and cGMP are ubiquitous second messengers that regulate many cellular functions including vision, sense of smell and taste, muscle contraction, neurotransmission, exocytosis and cell growth. Their concentrations are tightly regulated at the point of degradation by a family of enzymes called phosphodiesterases, which catalyse hydrolysis of cAMP and cGMP to the 5N-nucleotides AMP and GMP, Equations 3 and 4. Two metal ions, Zn^{2+} and Mg^{2+} are located $\sim 3.9\text{Å}$ apart at the base of a deep pocket in the enzyme and constitute a binuclear motif similar to that in the ribozyme above.

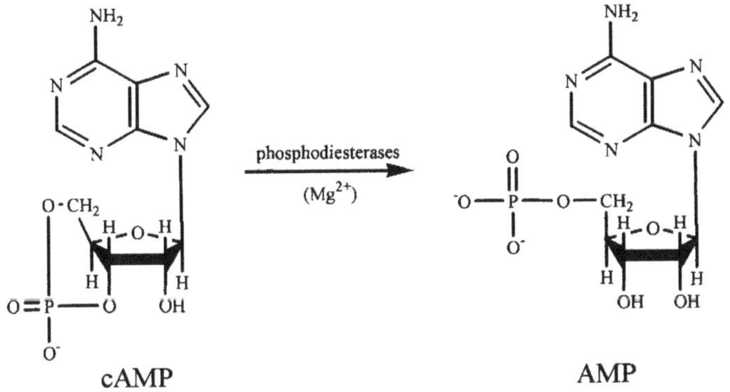

(3)

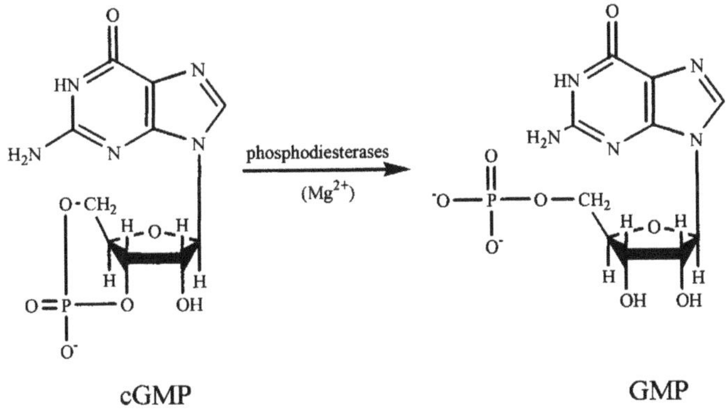

(4)

Not all enzymes that are activated by Mg^{2+} ions involve phosphoryl transfer, although this group is by far the largest, encompassing kinases, synthetases and phosphatases. In other Mg^{2+} ion activated enzymes, carboxylate binding to the magnesium ion is involved. Isoprenoids as a class include a vast number of natural products. They act as visual pigments, hormones and other signalling molecules and are components of cell membranes (*e.g.* cholesterol). Linear polyunsaturated molecules are converted to cyclic terpenes in complex, enzymatically catalysed cyclisation reactions, *e.g.* the conversion of farnesyl diphosphate to pentalenene, Equation 5.

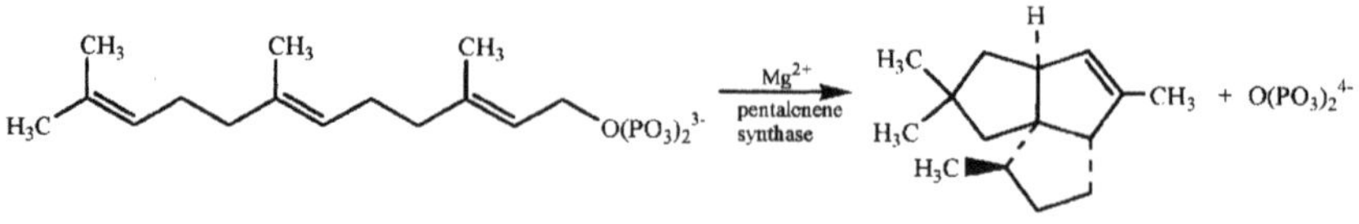

The structure of pentalenene synthase shows an aspartic acid-rich segment that forms part of the active site and provides carboxylate groups to which the magnesium ion coordinates. The Mg^{2+} ion probably stabilizes binding of the diphosphate groups in the substrate and is required to facilitate pyrophosphate release in the first step of the cyclization reaction. An aspartate(D)-rich DDXXD sequence thought to be the Mg^{2+} binding site is highly conserved in a number of terpenoid cyclases.

2.3.5 Photosynthesis

Magnesium strays from its propensity to complex with O-donor atoms in one important biochelate, the photosensitive centre in photosynthesis. Chlorins use the Mg^{2+} ion as the metal centre of choice. This is probably because, being redox stable, it does not interfere with photometric and free radical reactions that are the basis of the action of chlorophyll. Protoporphyrin IX is the last common intermediate precursor before branching occurs to go on to either chlorophyll or heme biosynthesis. Two different enzymes, Mg chelatase and Fe chelatase respectively, direct and catalyse these two branches. Magnesium chelatase (a complex enzyme with three protein subunits) catalyses the ATP dependent insertion of Mg^{2+} ion into protoporphyrin IX. This is a complicated two-step process involving preactivation and chelation, both of which are Mg^{2+} ion and ATP dependent.

2.4 CALCIUM

Calcium ions play roles ranging from minor to dominant in many important biological processes. Elaboration on many of those outlined here is provided in other parts of the text.

2.4.1 Impulse Transmission

The free Ca^{2+} ion is involved in charge movement in neuronal processes (4.4) and cardiac muscle. As important, however, is its ability to act as a cellular signalling device. In these respects it differs from the Mg^{2+} ion.

2.4.2 Second Messenger Action

Ca^{2+} is the only metal ion capable of acting as a second messenger and thus facilitating a variety of cellular processes. It is probably the most versatile of a limited number of second messengers, *i.e.* cAMP, cGMP, DAG, IP_3 and the Ca^{2+} ion. Their action is triggered by primary extracellular signals (first messengers) like hydrophilic hormones, neurotransmitters, growth factor, odours, light, *etc.* A diversity of events results that will be encountered subsequently. Calcium ion as a signalling device is all-pervasive.

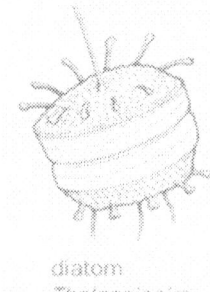

diatom
Thalassiosira

Calcium Signalling is Employed by the Simplest of Creatures

Diatoms are a group of unicellular algae with thin double shells of silica that look like a brown pillbox. They are the primary photosynthetic producers of organic matter in the oceans. Dead, they form certain sea floor rocks from which diatomaceous earth is obtained for industrial use as abrasives and filters.

The question arises – can such simple creatures respond to environmental changes? The answer is yes and calcium plays an important role in doing just this. Using transgenic marine diatom *Phaeodactylum tricornutum* cells containing aequorin (1.5.2), it has been shown that various imposed perturbations lead to transient increases in cytosolic $[Ca^{2+}]$. For example, shear (fluid motion) and osmotic stress (dilution) produced changes within seconds and addition of the nutrient Fe(III) ion produced a response after about five minutes that lasted for some time (1–2 hours). Calcium ion as a signalling agent is thus all-pervasive. The changes in $[Ca^{2+}]_{cyto}$ occasioned by the stimuli applied to the diatoms mirror changes observed in higher animals and which are characteristic of the activation of signal transduction.

2.5 CALCIUM BINDING PROTEINS AND THEIR OCCURENCE

Many of the actions of the Ca^{2+} ion take place because of its ability to bind strongly to biomolecules somewhere in the process. Calcium binding proteins can be broadly classified into strong binders with binding constants of $K \sim 10^6$ M^{-1} and weaker binders with K values of $\sim 10^3$ M^{-1}. Since the concentration of the Ca^{2+} ion is so much lower inside than outside the cell, the stronger binders are generally found inside the cell. The extent of binding of Ca^{2+} to these proteins can therefore change significantly when $[Ca^{2+}]_i$ changes from 100 nM to 1–10 μM, as it does when signalling occurs. Such trigger proteins, which are numerous, change their conformations when bound to Ca^{2+} ions and as a result modulate enzymes and ion channels. Strong binding proteins can also act as buffers by controlling the Ca^{2+} ion concentration within a cell. Weaker Ca^{2+} ion binding proteins are usually found outside the cell where the higher $[Ca^{2+}]$, typically mM, allows substantial formation of the Ca^{2+}–protein complex. These proteins can therefore act as Ca^{2+} ion buffers, *e.g.* calsequestrin, which has many calcium binding sites. Several calcium binding proteins and their functions are shown in Table 2.1.

2.5.1 Calmodulin

One of the most important and abundant calcium binding proteins is calmodulin (1.3.3), a protein that is present in all eukaryotic tissues. At

Table 2.1 *Selection of calcium binding proteins and their functions*

Protein	Function
Strong Ca^{2+} ion binders	
*Calmodulin	Targets variety of enzymes and proteins (Figure 2.3).
*Parvalbumin	Calcium buffer in muscle cells (particularly fish). Facilitates transfer of Ca^{2+} ions from myofibril to SR.
*Calbindins	Intracellular transport of Ca^{2+} ions, and buffer. Uncertain physiological function.
*Protein kinase C	Phosphorylating enzyme; activity greatly enhanced by Ca^{2+} ion.
*Troponin C	Mediates striated muscle contraction.
Annexin family	Membrane binding proteins involved in endo- and exocytosis. Also have anticoagulation and antiinflammatory properties. Uncertain physiological function.
Weaker Ca^{2+} ion binders	
Phospholipases	Hydrolysis of membrane phospholipids, *e.g.* phospholipase C (Figure 3.6). Most isozymes use Ca^{2+} ion at active site.
α-Amylase	Hydrolysis of α-1,4 link in glucose polysaccharides. Ca^{2+} ion near active site. Found in pancreatic juice and saliva.
Trypsin and serine proteases	Proteolytic digestive enzymes in pancreatic juice of mammals (structural Ca^{2+} ion).
Calsequestrin	Very acidic protein that binds many Ca^{2+} ions in SR of muscle cells.
Prothrombin	Blood clotting protein (Factor II) containing many gla residues (2.5.5).

* These are members of the calmodulin family and all contain E-F hands.

resting (low, $< \mu M$) $[Ca^{2+}]_i$ most of the calmodulin exists as the free ligand. An increase in $[Ca^{2+}]_i$ promotes the formation of the calcium–calmodulin complex (Equation 6).

$$\text{calmodulin} + 4Ca^{2+} \rightleftharpoons \text{calmodulin}(Ca^{2+})_4 \qquad (6)$$

The Ca^{2+} ion binding to calmodulin probably exposes a hydrophobic surface on the protein as a result of the induced conformational change. This, in turn, allows it to bind to certain enzymes and channels, thus activating them.

Ca^{2+}–calmodulin is essential for the successful action of a number of enzymes in the cytosol (Figure 2.3). These Ca^{2+}–calmodulin activated proteins often have a regulatory subunit to which the Ca^{2+}–calmodulin binds. Ca^{2+}–calmodulin also modulates certain ion channels and pumps, gene transcription, smooth muscle contraction (6.5.1) and activates membrane fusion. It is also important in signal transduction.

Only a Small Part of *Bacillus Anthracis* Makes it a Killer

Anthrax (Gk. coal, carbuncle) is a disease of sheep and cattle that can be transmitted to humans. *Bacillus anthracis* would be a harmless bacterium but for a small component that creates a protective capsule for the bacterium and another small portion that provides three proteins

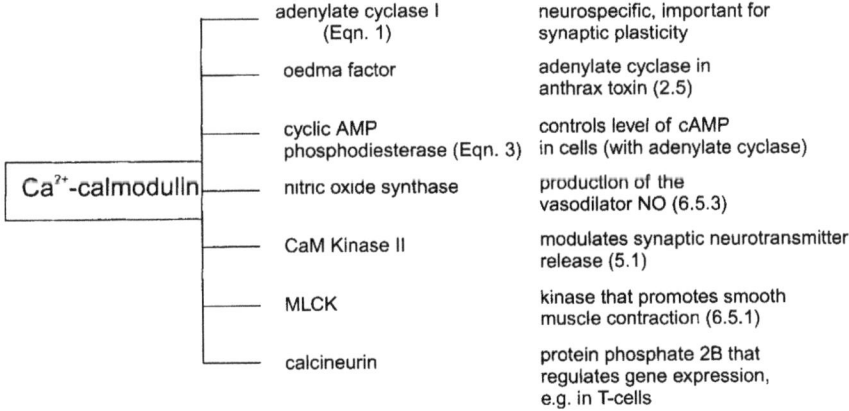

Ca²⁺-calmodulin		
	adenylate cyclase I (Eqn. 1)	neurospecific, important for synaptic plasticity
	oedma factor	adenylate cyclase in anthrax toxin (2.5)
	cyclic AMP phosphodiesterase (Eqn. 3)	controls level of cAMP in cells (with adenylate cyclase)
	nitric oxide synthase	production of the vasodilator NO (6.5.3)
	CaM Kinase II	modulates synaptic neurotransmitter release (5.1)
	MLCK	kinase that promotes smooth muscle contraction (6.5.1)
	calcineurin	protein phosphate 2B that regulates gene expression, e.g. in T-cells

Figure 2.3 *A variety of enzymes activated by Ca²⁺-calmodulin*

collectively referred to as anthrax toxin. All appear to be necessary to help inhaled *B. anthracis* spores to germinate in, and eventually kill, macrophages, the vital defensive immune cells in humans. In this most serious type of anthrax infection, the bacterium can then find its way into the bloodstream where it releases the three-component anthrax toxin.

The crystal structures of all three proteins have been solved. Lethal factor (LF) is a zinc protease that disrupts the signalling involved in the release of defensive proteins. Edema factor (EF) causes cell death and the accumulation of fluid (edema). Protective antigen (PA) binds to cell receptors and aids in the entry of LF and EF into the cell.

Calmodulin plays an integral role in the action of EF. The edema factor latches on to Ca^{2+}–calmodulin and changes the conformation of the latter so that it cannot bind to its intracellular targets. Thus, mediation of calcium signalling has been interrupted. Amazingly, at the same time the Ca^{2+}–calmodulin binding to EF modifies another region in EF to create a binding site for ATP. Now EF can act as an adenylate cyclase (an Mg^{2+} ion is also required) catalysing the conversion of ATP to cAMP (Equation 1). Increased (reaching 10^3-fold) levels of cAMP disrupt intracellular signalling pathways. In the adduct with EF, calmodulin binds only two Ca^{2+} ions at the 'high affinity' C-terminal lobe (1.3.3). Enzyme activation by calmodulin usually requires at least three bound Ca^{2+} ions. This could account for the slightly reduced effectiveness of Ca^{2+} ions and that calmodulin alone can still weakly activate EF. *Bordetella pertussis* (whooping cough) is another bacterium that causes an increase in cAMP by producing an exotoxin with calmodulin activated adenylate cyclase activity.

2.5.2 Lectins

Lectins are a diverse group of proteins that bind to specific sugars and play an important role in biological recognition. Several classes of lectins require Ca^{2+} ions to function. Although more often encountered in plants (where their specific function is still uncertain), lectins are known to be present in various mammalian tissues.

In the legume lectins, some of the Ca^{2+} ion ligands help to set up the sugar binding site using van der Waals interactions and hydrogen binding. In contrast, the calcium-dependent (C-type) animal lectins contain a Ca^{2+} ion that binds *directly* to two of the vicinal hydroxyl groups of the sugar. In addition, coordinated amino acids form a bridge between the Ca^{2+} ion and the hydroxyl groups. In rat (serum) mannose-binding protein A, the

Ca^{2+} ligands are glu-185, asn-187, glu-193, asn-205 and asp-206 as well as the C-3 and C-4 hydroxyl groups of mannose. This protein is one of the soluble collectins that bind to the surfaces of potential bacterial and viral pathogens and initiates the steps leading to the neutralization of the invading microorganism. Antifreeze proteins that lower the freezing points of fluids in certain teleost fish *e.g.* Atlantic herring, are homologues of the carbohydrate receptor domain of Ca^{2+}-dependent lectins.

2.5.3 Regulation of Gene Expression

Both Ca^{2+}–calmodulin and Ca^{2+} ions can modulate the activities of transcription factors.

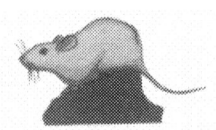

Dream on With Calcium

Calcium in the form of the Ca^{2+}–calmodulin protein complex can modulate the activities of transcription factors by indirect means involving protein kinases or phosphatases. An entirely new mechanism for the regulation of gene expression involving Ca^{2+} ions has recently been described. Transcription is repressed when the gene encoding human prodynorphin (which is involved in memory acquisition and in the control of pain) is tightly bound by the protein DREAM (Downstream Regulatory Element Antagonist Modulator). Prodynorphin is a precursor of dynorphin, a physiological neuropeptide that binds at opioid receptors in the brain and gut (7.7.1). Increased $[Ca^{2+}]$ in response to a cellular signal will bind to DREAM (at four E-F hand sites). The protein then dissociates from the DNA as a result of the structural changes caused by Ca^{2+} ion binding and the neuropeptide is now expressed.

Since DREAM protein suppresses the production of dynorphin, mice lacking DREAM overproduce this opioid in spinal cord nerve cells and are less sensitive to all varieties of pain.

2.5.4 Cell Birth and Death

In cell birth, calcium ions are essential to the function of sperm. It has been shown that there is a transient rise in $[Ca^{2+}]_i$ during the fertilization of the eggs of both vertebrates and invertebrates. The increase in calcium ion concentration in sea urchin eggs 15–35 seconds after the injection of sperm was observed by using a fluorescent dye as the Ca^{2+} indicator (1.5.2). After this time the egg knows that it can begin to divide. The binding of sperm to its plasma membrane receptor is a signal for the increase in [Ca^{2+}] in the egg cytoplasm to occur. This takes place through mediation *via* IP$_3$, produced by stimulation of hydrolysis of PIP$_2$ (Figure 3.6). This is demonstrated by blocking phospholipase C with an amino-steroid inhibitor, which retards the synthesis of IP$_3$ and suppresses sperm-induced Ca^{2+} ion release. In addition, a mouse sperm-specific phospholipase C isoenzyme triggers [Ca^{2+}] oscillations when microinjected into mouse eggs. These mimic those observed in normally fertilized eggs. The increase of [Ca^{2+}]$_i$ induces exocytosis of secretory vesicles beneath the plasma membrane of the egg, which probably alters the surface characteristics and prevents additional sperm from entering the egg!

A Calcium Channel and Sluggish Sperm

Mutant mice that have no gene encoding an ion (probably Ca) channel that is *sperm specific* become infertile. The sperm from the mutant mice can swim only about a third as fast as normal sperm and are unable to batter their way into an egg! A drug that temporarily blocks the corresponding calcium ion channel in human sperm could be an effective contraceptive and not affect other Ca channels in the body

The Human Sperm

In cell death, ions, particularly Ca^{2+}, have a major role. There are basically two kinds of cell death, although their distinction is somewhat blurred. In orderly or programmed cell death, *apoptosis*, the cell shrinks and pulls away from its neighbour. It is necessary for cell development as, for example, in tail elimination from tadpoles on becoming frogs. Changes in [Ca^{2+}]$_i$ are associated with the first step in apoptosis. What occurs then is still uncertain, but signal networks and catabolic enzymes could be activated. There is also accidental cell death, *necrosis*, caused for example by a stroke or a physical blow. Swelling and rupture of the damaged cell ensues, inhibiting pumping of Ca^{2+} ions, as well as Na^+ ions and H_2O, which now stream into the cell causing its death by an uncertain mechanism. A high influx of Ca^{2+} ions is associated with damaging cell death in Duchenne muscular dystrophy (6.2).

Spiders Cause Skin and Muscle Necrosis in Humans and Laboratory Animals

Skin necrosis (dermonecrosis) without damage to underlying skeletal muscle results from the gentle bite of some spiders of the genus *Loxosceles, e.g. L. recluse* (the brown recluse spider). Lesions from a bite are frequently large and can persist for weeks. Venom from certain tarantulas, *e.g. Duge-siella hentzi* (found in Arkansas) causes necrosis of skeletal muscle (myonecrosis) in mice. Rupture of the plasma membrane of the muscle cell and the inability of the mitochondria and SR to maintain normal (low) Ca^{2+} ion levels leads to cell death. Very high levels of Ca^{2+} ion were found in the necrotic muscle cells of the mice.

The protein p53 is a tumour suppressant. It signals cells to commit suicide (apotosis) when growth patterns run amok, as in cancer, for example. Dysfunctional p53 is involved in more than 50% of solid tumours. S100B protein is activated by Ca^{2+} ions inducing a large conformational change that is required for it to bind to p53. The binding results in a number of undesirable modifications in the actions of p53. Calcium ions and S100B thus regulate cellular functions of the tumour suppressor. S100B is implicated in a number of diseases and is a member of the E-F hand calcium binding protein superfamily (2.5).

2.5.5 Exocytosis and Endocytosis

In *exocytosis*, under the influence of Ca^{2+} ions, an intracellular vesicle fuses with the plasma membrane, the vesicle opens and its contents (neurotransmitters from nerve cells, hormones from endocrine cells,

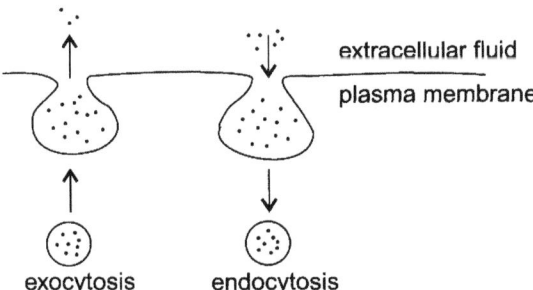

Figure 2.4 *Exocytosis and endocytosis*

neurohormones from nerve endings, digestive enzymes and even mucus (1.3.2)), are liberated into the extracellular fluid. *Endocytosis* is a similar process in the reverse direction. The plasma membrane folds into the cell and forms small pockets that break off to give intracellular membrane-bound vesicles (Figure 2.4). This has been much more difficult to study, but Ca^{2+} ions are also involved (5.1.1).

The elevation of $[Ca^{2+}]_i$ triggers the fusion of the synaptic vesicle with the plasma membrane. Just how this occurs is still unclear. It may be that *synaptotagmin*, a synaptic vesicle protein, binds to *neurexin* in the plasma membrane to form a fusion complex under the influence of Ca^{2+} ions. This docking is an important step in the release of the vesicle contents. Synaptotagmin contains three Ca^{2+} ions coordinated primarily by five aspartate residues located on two different loops. This is apparently a widely used C_2 motif (1.3.3). Lead ions can replace Ca^{2+} ions in synaptotagamin but do *not* promote the formation of the fusion complex. This could account for the disastrous effects of exposure to excess lead on mental and physical behaviour.

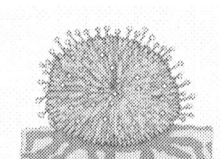

Waste Not, Want Not

Direct observation of the fertilization of sea urchin eggs (see cell birth), using a confocal microscope, shows a wave of secretory granule exocytosis followed rapidly by a wave of endocytosis. The granule membranes contain P-type voltage sensitive calcium channels. These are incorporated into the plasma membrane during exocytosis. The resulting localized entry of Ca^{2+} ions stimulates the endocytosis burst that specifically retrieves granule membrane compounds for reuse. Sites of exocyotosis and endocytosis are therefore confined to the same area.

2.5.6 Movement in Organisms

There are different types of movement of, and within, organisms and in these, Ca^{2+} ions play a pivotal role (aided by Na^+, K^+ and Mg^{2+} ions at various stages). They include ameboid locomotion (6.5.4), movement or tension in a body from muscle contractions of all types (Chapter 6) and cilium (*e.g.* in paramecium, 7.2.1) or flagellum (*e.g.* in sperm, 2.5.2) promoted motion.

2.5.7 Blood Coagulation

In part this is a tale of calcium binding proteins. When blood vessels rupture, the process of blood coagulation is quickly activated. It is a very complex procedure using a many-step cascade of proteolytic reactions and numerous feedbacks. Calcium ions in the plasma are required in many of the steps. Fortunately there are always sufficient Ca^{2+} ions available (otherwise muscle contraction would cease and death due to cardiac failure would result!). A number of blood proteins, called coagulation factors, require Ca^{2+} ions for activity. The γ-carboxyglutamyl residues, gla (1), present in all these factors bind to Ca^{2+} and this enables the factor to be anchored to the phospholipid membrane *via* calcium ions acting as links.

In the final step, the coagulation factors X_a and V_a ($_a$ = active form) in

$$
\begin{array}{cc}
\underset{\substack{\| \\ O}}{\overset{}{-C}} - \underset{\substack{| \\ NH}}{CHCH_2CH(CO_2^-)_2} &
\begin{array}{c} N(CH_2COOH)_2 \\ / \\ (CH_2)_2 \\ \backslash \\ N(CH_2COOH)_2 \end{array} \\
(1) & (2)
\end{array}
$$

(1)(2) *In gla-containing proteins as many as 9–12 gla residues (1) can be present. These proteins are found in bone and matrix material as well as in peptides in Conus venom (Table 5.3). Osteocalcin is a bone-specific protein containing three gla residues, which may induce mineralization in bone (8.4.2)*

conjunction with Ca^{2+} ions in the membrane convert prothrombin to thrombin (the effect of X_a alone is accelerated by 10^4 when V_a and Ca^{2+} ions are added). Thrombin catalyses the conversion of fibrinogen (a soluble plasma protein) to fibrin, first as a soft clot, and then under the influence of factor XIII_a converts to a fibrin (hard clot) that is a polymeric fibrous network. The hard clot enmeshes red blood cells and in this way acts as a plug to prevent serious blood loss. Blood platelets also contain actomyosin, which causes them to contract when blood clotting is underway, thus strengthening the clot (Ca^{2+}–calmodulin activation of MLCK in smooth muscle, Figure 6.14). The importance of calcium to the clotting process is clearly demonstrated by the addition of EDTA (2) to fresh blood (*e.g.* at a homicide scene) to sequester Ca^{2+} ions and prevent clotting and thus help in the analysis of the blood sample.

Rodent Killer is an Anticoagulant!

The importance of the gla residues in a number of coagulation factors is shown by the use (for over 50 years) of warfarin (3) as an oral anticoagulant (for blood clot dissolution) and as a rodent killer! The drug interferes with the action of vitamin K, which is a cofactor in the γ-carboxylation of glu to gla residues and therefore in the synthesis of prothrombin and other coagulation factors.

(3)

(3) *The commercial product is the racemic mixture, but it is the S(–) enantiomer that is the more active*

2.5.8 Biomineralization

Nearly 30 different minerals that appear in biomineralization have Ca^{2+} as their major cation. Other metal ions are sometimes minor contaminants. Many organisms can convert ions to solid minerals that fulfil many functions. Coordination of Ca^{2+} ions to biomolecules controls the steps that lead to mineral deposition and, remarkably, the nature of the calcium mineral.

Just 1% Addition Makes All the Difference!

Aragonite is a brittle crystalline material. However the addition of only 1% of a protein 'glue', consisting of polysaccharide and protein fibres laid down by the cellular mechanism of the giant pink queen conch (*Strombus gigas*), produces a shell a thousand times tougher than one of pure aragonite crystals.

Layers of tiny platelets of the mineral held together by protein are arranged at right angles to one another. The platelets dissipate stress into many cracks and thereby strengthen the composite.

2.6 CLINICAL ASPECTS

2.6.1 Dietary Requirements

The dietary requirements of the essential s-block elements for a normal adult are shown in Table 2.2. However care must be taken, particularly over the consumption of Na^+ ions!

In Humans, Too Much Na^+ Ion Can Be a Bad Thing, or Drinka Pinta Beera Day

A controversy has arisen over the possible dangers of excessive Na^+ ion in the human diet. Human beings are unique amongst vertebrates in the tendency to develop high blood pressure. Hypertension contributes to heart diseases, strokes and kidney failure.

Recommended treatments for the control of hypertension include restricting the ingestion of sodium chloride (salt) or, if necessary, by medication (Ca^{2+} ion channel blockers). Although older hypertensive individuals probably benefit from limiting their sodium intake, it has been suggested that the emphasis on Na^+ ion as the only dietary villain is too narrow. A diet low in fat and high in Ca^{2+} ion and that includes plenty of fruits and vegetables has also been shown to be effective in controlling hypertension (and, incidentally in reducing osteoporosis and certain cancers). New US government sponsored research does indicate however that for all individuals, blood pressure falls with a low sodium diet. It is interesting to note that the Na^+:K^+ ratio is 1:4 in beer compared to a 30:1 ratio in some popular sports drinks. Those who drink a pint of beer a day are at less risk from heart disease than those who drink heavily or not at all.

2.6.2 Deficiencies and Excesses

Ordinarily a healthy person on a sensible diet experiences none of the problems that can occur when metal ion concentrations in the body are above or below their normal concentrations. Problems that arise because of an excess or deficiency of a metal ion are shown in Table 2.3.

Deficiencies can arise from alcohol abuse, the use of diuretics, sustained diarrhoea, burns and renal diseases. Addition of the element either orally

Table 2.2 *Dietary requirements and distribution in humans*

Element( )[a]	Important Source	Dietary Requirement (adult, daily in grams)	Distribution
Na (1940)[b]	Salt added to food.	1.1–3.3	About 40% in bones and teeth, remainder mainly in extracellular fluids.
K (1926)	Citrus fruits and juices, vegetables,[c] nuts, cereals, chocolate.	1.9–5.6	Major (90%) intracellular cation. Less than 8% in bones.
Mg (1932)	Green vegetables, nuts, legumes, bananas, chocolate.	0.35	Least abundant of the four elements, 60% in bones and teeth.
Ca (1842)	Milk and other dairy products, beans vegetables.	0.80	99% in bones and teeth, remainder free ion or bound to plasma proteins.

[a] Year when first shown to be essential. [b] Need shown in cattle 100 years earlier. [c] Most plant cells accumulate K^+ and reject Na^+ ions and therefore are a good source of potassium.

Table 2.3 *Clinical problems arising from deficiencies and excesses of metal ions*

Element	Deficiency	Excess
Na	Hyponatremia, serum $[Na^+] < 136$ mM, caused by H_2O shift from extra to intracellular and cells swell. Decreased blood pressure, muscle spasms and coma.	Hypernatremia, serum $[Na^+] > 145$ mM, caused by excessive loss of H_2O from cells. Hypertension, mental confusion and coma.
K	Hypokalemia, serum $[K^+] < 3.5$ mM, abnormal ECG. Muscle problems, cardiac problems, mental confusion, and coma.	Hyperkalemia, serum $[K^+] > 5$ mM. Excess is toxic (lethal dose, 14 g). Neuromuscular and metabolic problems; life threatening arrhythmias.
Mg	Hypomagnesemia, (2.1.2) serum $[Mg^{2+}] < 1.3$ mM, rare. Neuromuscular problems, seizures. Wasting syndrome with excessive loss (also Ca^{2+}) leading to kidney failure and mental and cardiac problems.	Hypermagnesemia, serum $[Mg^{2+}] > 2.1$ mM. Symptoms similar to those for hyperkalemia.
Ca	Hypocalcemia, total serum $[Ca^{2+}] < 2.1$ mM. Muscle spasms, irregular heartbeat. See: rickets (vitamin D deficiency) and osteoporosis (8.4.2). Poor blood clotting, prolonged QT interval in ECG, hypocalcemic tetany (see skeletal muscle).	Hypercalcemia, serum $[Ca^{2+}] > 2.6$ mM, often from nutritional supplements. Metabolic interference (increased kidney stone formation), affects all muscle types (excess calcification of bone), neurological (mental disturbances), shortened QT interval in ECG.

or by injection is usually an effective treatment. An excessive intake of metal ions often arises from overzealous use of mineral supplements or from salt substitutes that are present in certain medications.

2.6.3 Medicines

Although sodium salts (*e.g.* NaCl) are often used as auxiliary material, only Mg^{2+} and Ca^{2+} ions feature as important components of medicines. The mineral Epsomite ($MgSO_4.7H_2O$) was first found in springs in Epsom in the seventeenth century. It is used in Epsom salts as a laxative. The heavily hydrated (many H_2Os of hydration) Mg^{2+} ion can carry lots of water through the intestinal tract. Stretch receptors there are stimulated and evacuation reflexes are promoted. Magnesium carbonate and mixtures with calcium carbonate are simple antacids. We only encounter lithium in this book in a clinical context.

Calming Effect of Lithium

It has been estimated that about one person in a thousand in highly industrialized countries undergoes Li therapy for the treatment of bipolar disorder. This is a mental disorder characterized by periods of mania and depression. Just over a gram a day of lithium carbonate or citrate stabilizes the cycle of mood changes within three weeks. Introduced into psychiatry half a century ago, the basis of its action is still unknown.

It has been suggested that Li^+ ions can compete with Mg^{2+} ions in an enzyme that is responsible for the transcription of a protein (MARCKS) gene. Excessive production of MARCKS, which may contribute to manic depression, is thereby prevented. As well as lithium, carbamazepine (4) and valproic acid (5) are used to control manic depression. It has been found recently that all three drugs block the production of inositol (6) in cultured nerve cells. Inositol is required to maintain the level of inositol lipids such as phosphatidylinositol-4,5-biphosphate (Figure 3.6). These lipids in turn are used to generate IP_3 and DAG, which are important signalling agents (Figure 3.6). Inositol phosphate metabolism is therefore an important area upon which to focus in understanding the action of these drugs.

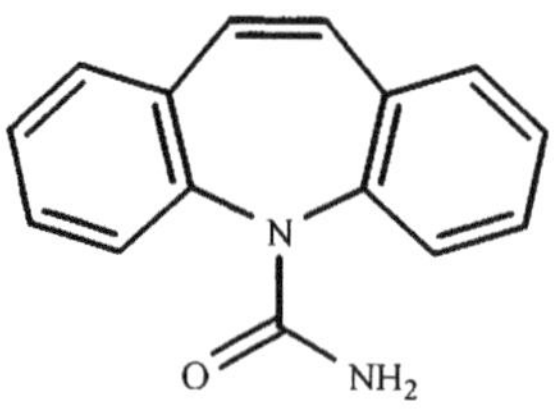

(4) 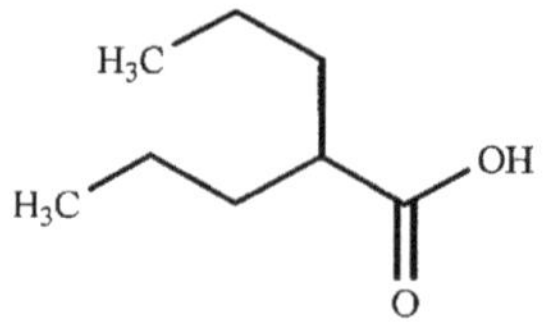   (5) 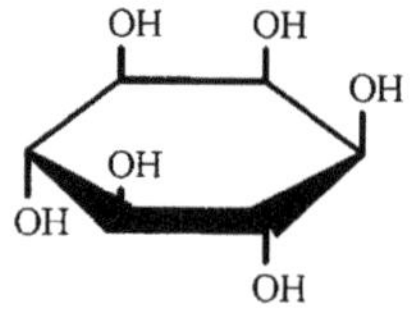   (6)

2.7 SUMMARY

The biological roles of the ions were outlined. A number of the functions are associated mainly with only one or two of the ions. These include solute transport (Na^+), enzyme activation (K^+ and Mg^{2+}) and interactions with biomolecules (Mg^{2+} and Ca^{2+}). Functions such as impulse transmission require the participation of the Na^+, K^+ Ca^{2+} and Cl^- ions. Remarkably, the effect of Ca^{2+} cannot be duplicated by any other ion in a number of activities such as second messenger action, assisting in blood coagulation or in biomineralization. Many of these roles are covered in detail in the remaining Chapters 3–8. Finally, the clinical aspects of these ions, their dietary requirements in humans and the dangers of a deficiency or of an excess were described.

Moving Ions Through Membranes

Prokaryotic and eukaryotic cells are enclosed in plasma membranes that prevent the exchange of material between the cell and the outside environment.

All membranes contain lipids and proteins. There are three major types of phospholipid asymmetrically distributed between the two halves of the bilayer, which is 50–70 Å thick. The phospholipid molecules are oriented with the fatty acid chains in the interior and the hydrophilic phospholipid head groups on the surfaces to form a lipid bilayer (Figure 3.1).

3.1 WHY AND WHERE CHANNELS OCCUR

It is necessary to move solutes across cell membranes of all types to accomplish many of the diverse processes that are required to maintain life. These include:

- Signalling in the nervous system.
- Coordination of muscle changes.
- Control of cell volume.
- Regulation of ion concentrations within cells (intracellular control of Ca^{2+} and H^+ are particularly important).

When ions are involved, the alien nature of the hydrophobic membrane does not allow unaided direct transfer even when there is a concentration gradient, and it is therefore necessary that ion channels are available. These are pores in the membrane, formed by proteins, which allow the movement of ions from one milieu to another.

Every membrane in all cells (nerve, muscle, epithelia, *etc.*) in all locations contains a variety of channels that are used to move H^+, Na^+, K^+, Mg^{2+}, Ca^{2+} and Cl^- ions. We will discuss H^+ and Cl^- ion channels when pertinent. These play a vital role in the control of pH and skeletal

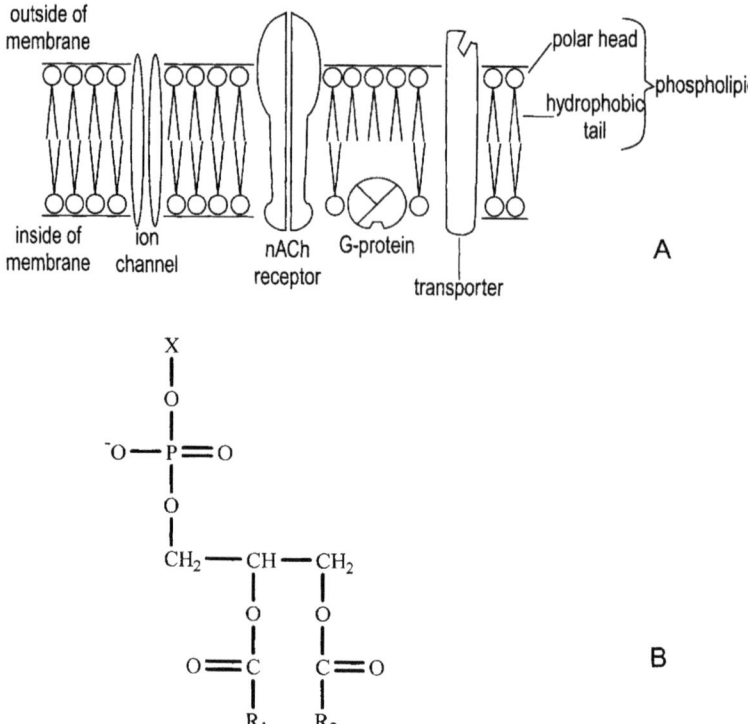

Figure 3.1A *Plasma membrane and some associated proteins (fluid mosaic model). The proteins are embedded within either the whole membrane (intrinsic) or only one half (leaflet) (extrinsic) and vary in the amount exposed to the aqueous environment on either side of the membrane. They are usually attached to the membrane via α-helical regions (20–25 hydrophobic amino acids) that can span the membrane one or more times. They are difficult to remove intact for examination. Only a dozen or so membrane protein structures have been solved by X-ray crystallography, compared with thousands of other protein structures. There are a variety of such proteins that function in a number of ways to transpose ions (and other solutes) from one side of the membrane to the other. Several different types are represented in the figure. In addition the animal plasma membrane contains an abundance of cholesterol.*

Figure 3.1B *Structure of glycerophospholipids (the major lipid components of membranes). X can be one of a variety of polar groups. R_2 and R_1 are saturated C_{16} and C_{18} and unsaturated C_{16} to C_{20} long chain aliphatics respectively. An example is PIP$_2$ (Figure 3.6).*

muscle function, as well as being involved in cell volume regulation and solute transport across epithelial surfaces.

Sometimes the channel is nonselective or only weakly selective towards ions (3.5.1), but often there is significant selectivity for a particular ion. For example, there can be a 10^3 preference for entry of K^+ over Na^+ through a certain channel, which is therefore termed a K channel. There are at least 30 types of K channels with different characteristics and functions. There are general similarities in structure and behaviour of specific ion channels obtained from disparate species. For instance, a certain type of Ca channel is found to be similar in yeast, worms and humans. The Na channel in the electroplates in electric eels resembles those in jellyfish, fruit flies and human nerve fibres. It is clear from this that the results of studies of channels from invertebrates will be *generally* applicable to vertebrates.

 **Many Similarities and Few Differences Between Nematode and Vertebrate Channels**

Caenorhabditis elegans is a tiny roundworm about 1 mm long with a translucent body that allows easy viewing of its neuron network. In 1998, the DNA sequence of the entire genome of the creature was mapped. The genome encodes about 80 K channels, 90 ligand gated channels, 6 cyclic nucleotide gated channels, 5 voltage gated Ca channels and 6 Cl channels. No voltage gated Na channel genes are present, consistent with the absence of electrophysiological activity associated with such channels. Voltage activated channels are abundant. We shall encounter all of these channels later.

Some channels are able to transfer ions at a remarkable rate, as much as 10^8 ions per second through each channel, thereby generating currents as small as 10^{-10} amps per channel. It has been a remarkable development in recent years that this small current in an *isolated* single channel can be measured (patch clamp method) (Figure 4.7). Ion channels are not continually open, but respond to different stimuli. A variety of perturbations can lead to the opening or closing of an ion channel (called gating), which is usually brought about by a conformational change in a protein incorporating the channel.

3.2 CHANNEL STIMULI AND TYPES OF CHANNELS

There are a number of types of perturbations that can have a profound effect on the state (open or closed) of a channel. Stimuli might arise in the skin, muscle, bone or associated organs or be unconsciously controlled in internal organs. The sensory cell activity both detects and then transduces the stimulus, in all cases altering the activity of an ion channel directly or indirectly. The perturbations that are likely to be encountered in most animals include:

- *Mechanical stresses.* These include pressure, touch and the special example of sound (in hearing). The receptor protein incorporates an ion channel that opens directly on stimulation (physical pressure) *via* a conformational change. Mechanical sensors are present in a wide range of creatures (7.2).
- *Temperature changes.* The state of certain cation channels (open or closed) can be directly affected by temperature changes. The channels discriminate poorly among alkali metal ions.
- *Light.* The conversion of a visual signal to an electrical one using the usual ion channels is probably the best understood of all sensory mechanisms.
- *Chemicals.* These can originate externally and be the reason for example, for odour or taste responses. They can also be generated internally. These stimuli and their management are treated in Chapter 7.

Stimuli are processed by two classes of channels that respond to:

- *Membrane voltage changes.* These operate on voltage gated channels. The terms voltage activated, voltage operated and voltage dependent are also used. The opening (and closing) of the channel is a direct result of a voltage change and tends to be ion-selective.
- *Ligand binding.* This occurs at a specific site on the so-called ligand gated channel or receptor. The ligand may arrive from an external source or be an internal 'messenger'. The effect is direct when the ligand binding site and the ion channel are part of the same molecule (ionotropic receptor). In some cases, the receptor (metabotropic) is linked indirectly to an ion flux. Ligand gated channels are not especially ion discriminating.

3.3 VOLTAGE GATED CHANNELS

Such channels play an especially vital role in conferring excitability and this allows the transmission of electric signals in nerve, muscle and sensory cells. The interplay of opening and closing the various sodium, potassium and calcium channels (as well as chloride channels) in membranes (gating) provides the necessary charge transfer along the nerve. These channels are mainly closed at the resting potential of the cell, but open in response to voltage changes across the membrane and exhibit varying degrees of voltage sensitivity. Some voltage-sensitive channels also respond to ligand binding and are subject to modulation by cytosolic substances like H^+, Ca^{2+}, ATP, *etc.* Voltage gated channel function can be analysed in terms of closed, open and inactivated states. It must be emphasized that these are the major conformations and that other substates also exist. Some important voltage sensitive channels that will be encountered later are shown in Table 3.1.

3.3.1 Sodium Channels

There is a high concentration of sodium channels in brain tissues. The electric organs of electric eels, genus *Electrophorus* and of electric rays, genus *Torpedo*, are a particularly good source and it was from these species that the first voltage gated channels were isolated. The channel blockers tetrodotoxin (TTX, 1) and saxitoxin (STX, 2) bind very specifically to Na channels. Detergent solubilized membrane fragments of *Electrophorus* were separated by affinity chromatography using bead-immobilized TTX. The α-subunit of the sodium channel of the eel electroplate on the electric organ has been cloned and sequenced using recombinant DNA techniques. Other smaller subunits (β) are much less functionally important.

Table 3.1 *Some voltage sensitive channels*

Name	Function
Na	In nerve and muscle fibres.
	Provide upstroke of action potential in neuron membranes and most electrically excitable cells.
	Blockers are TTX and STX (1) and (2).
K_V	Classical and main K channel.
	Activated by depolarization, responsible for repolarization, but slower response than K_A and Na channels. Blocker is $(C_2H_5)_4N^+$ (TEA).
K_A	Activated by depolarization after a period of hyperpolarization.
	Opens and switches off rapidly.
K_{Ca}	Responds to Ca^{2+} ions (μM) from stores and *via* voltage gated Ca channels. Key regulator of vascular smooth muscle.
	Only one subtype (BK, also called maxiK), also has moderate voltage dependence of activation.
	Blocker is charybdotoxin (Table 4.5).
	Weak voltage dependent K_{Ca} channels also exist (subtypes SK and IK).
Ca_L	In a wide variety of tissues.
	Maintains depolarization during action potential plateau in vertebrate cardiac muscle. L = long open times (compared with Na channels).
	Blocked by DHP (3).
Ca_N and Ca_P	Trigger release of neurotransmitter from presynaptic endings in birds (N-type) and mammals (P-type)

(1)   (2)

*(1)(2) TTX and STX are natural paralytic toxins that act as specific Na channel blockers
thus shutting down the nervous system. In contrast to synthetic Na channel blockers,
TTX and STX are unique in being voltage-insensitive. The hydroxyl and guanidinium
groups are important for activity. TTX and STX are unusual neurotoxins in that they
are not proteins (Table 4.3)*

Until recently, voltage gated ion channels were not thought to exist in
prokaryotes. However, now the voltage gated ion channel from the salt-
tolerant extremophile *Bacillus halodurans* (~1 M salt and pH ~11) has
been expressed and characterized. The results were surprising. Although it
is blocked by Ca channel blockers, it is selective for sodium, but not
blocked by TTX. It has a much simpler structure than the Na$_v$ channel of
eukaryotes in that there is a single domain rather than four (see below).

3.3.2 Potassium Channels

Potassium channels are widely distributed in the animal kingdom and
dominate in plants. Most K channels are voltage gated. They are many and
diverse in their roles. They are responsible for regulating the membrane
resting potential in excitable and nonexcitable (*e.g.* epithelial) cells. They
terminate action potentials and limit their firing frequencies (4.4). BK$_{Ca}$
channels are opened by depolarization, but the voltage dependence of
activation is shifted to more negative membrane potentials in the presence
of Ca^{2+} ions, *i.e.* they then open more easily. There are a large number of
genes that encode for voltage gated K channel-like subunits in the *C.
elegans* genome (3.1), the first metazoan genome to be completely
sequenced. It is unclear how the cell knows which gene to express in order
to take advantage of such a diverse number of K$_v$ channels and why so
many are required.

Rich sources of and strong specific binders to K channels (two
requirements for their isolation) are so far unavailable. The structural

examination of these channels has therefore relied on a different approach to that used for Na (and Ca) channels. Quite surprisingly this came first from examination of a mutant strain of fruit flies (*Drosophila melanogaster*) that lack one particular gene (*Shaker* locus).

Ether Shakes Up a Particular Fruit Fly

The *Shaker* voltage gated potassium channel is named after a fruit fly mutant that shakes its leg when anaesthetized with ether (a chance observation!). The defect in the K_A channel (which contributes substantially to repolarization in the axon fibres) results in the mutant form having abnormally long action potentials, prolonged transmitter release and therefore extended muscle contraction, shown by 'shaking'. Cloning of the genomic DNA from the *Shaker* locus allowed (with difficulty) the amino acid sequence of the putative potassium channel protein to be deduced. The sequence is similar to that found subsequently for many other K^+ channels.

3.3.3 Calcium Channels

Calcium channels play important roles in excitable and nonexcitable cells in organisms ranging from paramecia to man. They feature in neurotransmission, muscle contraction and secretion. Voltage gated Ca channels are entirely responsible for depolarization in most invertebrate muscles and vertebrate smooth muscles. They assist Na channels in vertebrate cardiac muscles. They are unique in transmitting a messenger (Ca^{2+}) ion. There are a number of major types, all voltage gated, distinguished by physiochemical properties for example, the magnitude of depolarization necessary to open the channel, and (so far) six letters of the alphabet (T, L, N, P, Q and R)! The Ca(L) channels are best understood. Those used in cardiac and smooth muscle are similar to those in skeletal muscle except for subtype differences. Even though these are wider channels than those for Na^+ and K^+ ions, they must be *very* selective against Na^+ and K^+ ions and are so in the presence of external Ca^{2+} ions. They are sensitive to a number of neurotransmitters and drugs, most of which inactivate the channel and are therefore called calcium antagonists. One of these, 1,4-dihydropyridine (DHP) (3) binds strongly only to the L-type channel and has been used to isolate this calcium channel from rabbit skeletal muscle. The N-, P-, Q- and R-types are blocked by tarantula venom. The peptide from the venom of the world's largest tarantula (Cameron Red Baboon) protects mice from seizures and is the first substance known to block R-channels, which remain shut until stimulated by a signal from a neighbouring neuron. They have been implicated in epilepsy.

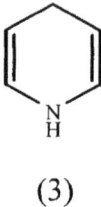

(3)

3.4 STRUCTURAL ASPECTS

There is a high degree of structural similarity among all the voltage gated ion channels irrespective of their source. All consist of two or more subunits, but only one, the principal (α-) subunit, is very important to the function of the channel. This is because it supplies the channel pore, is responsible for voltage sensing and is consequently a target of drugs, toxins and crippling mutations. The other subunits modulate channel properties. The rates of gating of Na channels are increased by the β_1 subunit and the amino terminus of a β subunit of the K channel provides the plug to the central cavity of the pore.

3.4.1 Topology of the α-Subunit of Voltage Gated Ion Channels

The topology (Figure 3.2) is deduced from knowledge of the hydrophobic and hydrophilic regions of the ion channel protein. It is surmised that the hydrophobic region is more likely to traverse the membrane.

Four domains, I–IV, are joined together in the voltage gated Na and Ca channels to form a single polypeptide chain of about 1600 amino acids, and this polypeptide folds to form a pore. Each domain consists of six transmembrane α-helices (S1–S6). The voltage gated K_v channel α-subunit consists of only six transmembrane α-helices (*i.e.* corresponding to region I), but four identical (homomeric) or slightly different (heteromeric) domains associate to form the functional channel. In all channels,

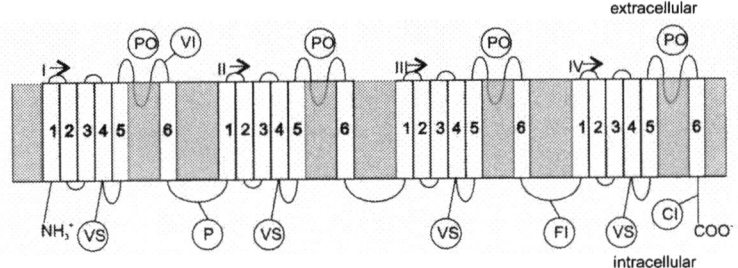

FI) fast inactivation (Na only) CI) calcium-dependent inactivation (Ca only) VI) voltage gated inactivation (Ca only)

P) phosphorylation sites (not always present) PO) contribution to channel pore VS) voltage sensor

Figure 3.2 *Topology of the α-subunit of voltage gated Na⁺ and Ca²⁺ ion channels and action sites. The cylinders labelled 1–6 represent transmembrane α-helices of about 20 nonpolar amino acids. This is a sufficient length to span the membrane (about 50 Å thick). Shorter sequences indicated by lines link the transmembrane helices and dip into and out of the membrane. Residues that link the domains, as well as the N and C termini, lie inside. There are four domains, I–IV, in Na and Ca channels, and only one domain in a K_V channel. The non-voltage gated KcsA channel contains only the 5 and 6 segments connected by a pore loop. Both the K_V and the KcsA channels form tetramers. The locations in the channel that are believed to be responsible for various features of the action potential (4.4 and 6.4.2), the sensitivity to voltage and the contribution to the channel pore are shown*

both the N- and C-termini are intracellular. The α-helical links are of variable lengths and the intracellular portions are usually longer.

In all cases the S5–pore–S6 segment lines much of the channel and contains the selectivity filter where the pore is narrowest and this is what determines which ion is chosen. This is well illustrated by the observation that site directed mutagenesis of amino acids (glutamate for lysine and alanine) in the pore region results in the conversion of a channel selective for Na⁺ ions into one selective for Ca²⁺ ions. We have previously described (Figure 1.1) the mode of action of the selectivity filter in the KcsA channel, and the ability of the pore to transmit the (larger) K⁺ ion rather than the Na⁺ ion.

The S4 helix provides the voltage-sensor unit. It has many more positively charged amino acid residues (arginines and lysines) than is usual. Site-directed mutagenesis has confirmed that positively charged arginines are involved in voltage gated Na channels. These react to transmembrane voltages (depolarization) by moving toward the extracellular side of the membrane (negatively charged) thus promoting conformational changes that open the channel and allow the flow of ions down their concentration gradient. Specific details are still lacking.

Voltage gated Na and K channels consist of α- and one or more β-subunits. The skeletal muscle L-calcium channel is more complex with α_1, α_2, β, γ and δ-subunits, of which the α_1 is the largest (190 kDa). All except the β-subunit have hydrophobic (membrane) domains.

Rocking and Rolling Mice and Calcium Channels

Different types of mutations in various subunits of ion channels lead to animal diseases (4.5.1 and 6.6). Known mutations in an α_{1A} subunit of a voltage gated Ca channel (one of at least 10 classes of α_1 subunits) cause seizures and neuronal degradation of varying intensity in humans and mice. These show up in three human diseases, including a vicious rare type of migraine with aura (familial hemiplagic migraine, Table 4.2 and Figure 6.17) and abnormal phenotypes in mice. Mutant mice called Tottering, Leaner, Rolling and Rocker (describing their behaviour) also express variants of the α_{1A} subunit. Mutations in genes encoding the β subunit of neuronal voltage gated Ca channels leads to lethargic mice.

'Knockout' mice, in which the α_{1A} subunit has been deleted, can be genetically engineered. These mice deteriorate rapidly after birth and exhibit ataxic behaviour reminiscent of that of the wild type mutants. They completely lack P and Q type voltage gated Ca channels. This 'knockout' approach provides new insights into channel functions and augments classical studies using channel targeted drugs.

3.4.2 The KcsA Channel Pore

The cloning and functional expression of the first prokaryotic gene encoding a potassium ion channel (KcsA) from *Streptomyces lividans* enabled sufficient quantities of protein to be obtained for X-ray crystallographic analysis of KcsA. This channel differs from the voltage gated K channel in that it is not voltage activated, it has only two transmembrane segments with their connecting loop (corresponding to S5–pore–S6) and it lacks the voltage sensing S4 helix. Four domains are still joined together to form the pore (Figure 3.3).

The question arises, of course, as to whether the channel details obtained for a non-neuronal bacterium can be extrapolated to those of a neuronal potassium channel of a eukaryotic cell (structural details for which are still awaited). The answer is probably yes! For example, scorpion toxin binding patterns to both types of channel are similar. In

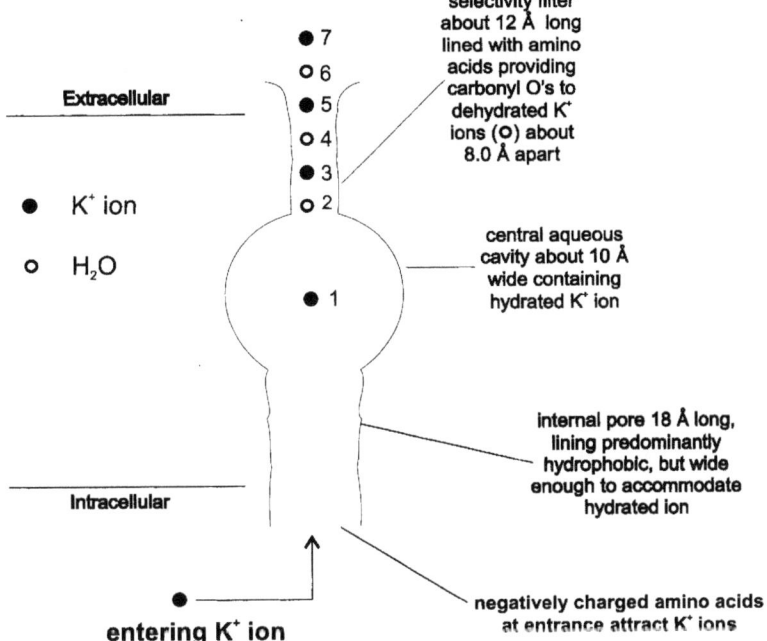

Figure 3.3 *Diagramatic representation of the pore of the KcsA channel. There are seven distinct sites (1–7) for K⁺ ions along the pore. The site 1 in the central cavity is fully occupied by an aquated K⁺ ion. The remaining six sites, 2–7, are half occupied in an alternating fashion at any one time. Four sites, 2–5, are in the selectivity filter and two sites, 6 and 7, are just beyond the end of the pore, probably held there by the channel surface. A K⁺ ion entering from the interior moves the hydrated K⁺ ion from site 1 to site 2, in which process the waters of hydration are replaced by oxygen atoms from the protein (Figure 1.1). Potassium ions are then moved stepwise until the one at the end (site 7) is displaced. Although the K⁺ ions are tightly bound in the selectivity filter, the two bound potassium ions repel one another and allow speedy movement. The side chains lining the central cavity help to maintain the primary water molecules around the K⁺ ion via secondary waters and weak hydrogen bonds. This allowed the structure of the aquated K⁺ ion held in the centre of the cavity to be determined. It is K(H₂O)₈⁺ with the square antiprismatic arrangement of H₂Os that is similar to the structure of the K⁺ ion within the selectivity filter. This reduces any barrier to the movement of the K⁺ ions*

Reprinted with permission from *Nature* from Y. Zhou, J. H. Morales-Cabral, A. Kaufman and R. MacKinnon, 2001, 414, 43–48, copyright 2001 Macmillan Publishers Ltd.

addition, within the pore region, the amino acid sequences of KcsA and K_V are nearly identical.

Both the 2-transmembrane and 6-transmembrane K⁺ ion channels are found in humans. The former, (K_{ir}) are responsible for the regulation of the resting membrane and for K⁺ transport across membranes. They are voltage-independent, but open at membrane potentials near or more negative than the resting potential and then allow K^+ ions to flow *into* the cell. They are modulated by agonists and antagonists (*e.g.* ATP, see 3.5). The 6-transmembrane channels are voltage gated and are employed in nerve and muscle fibres and endocrine cells.

3.5 LIGAND PROMOTED CHANNEL OPENING

Ligand binding to a receptor, rather than a membrane voltage change, can also produce an ion channel response. If the receptor incorporates an ion channel (ionotropic receptor) ligand binding usually results in stabilization of the open channel state, although a few ligands (such as ATP) promote channel closure. Such K_{ATP} channels are widely distributed in vertebrate tissues. They close rapidly when ATP binds and delay opening when ATP is released from the channel. Unlike other ATP-dependent channels (3.6.1) the hydrolysis of ATP is not involved. The K_{ATP} and a voltage gated Ca channel are important in the production and secretion of insulin.

Few Would Refuse Chocolate Given as Medicine

When the level of blood glucose is low ($< 3\,\text{mM}$), K_{ATP} channels in β-pancreatic cells (involved with production and secretion of insulin) are open. K^+ ions can flow out of the cell (Table 1.1) leaving the membrane potential quite negative (hyperpolarized) (Chapter 4). When the blood glucose concentration rises, for example after a meal, glucose passes through a transporter and its metabolism produces ATP. The [ATP] increases in the cell and it binds directly to the K_{ATP} channel and closes it. The membrane potential becomes less negative and this is a signal for a Ca channel associated with the membrane to open. Ca^{2+} ions flow into the cell (Table 1.1) and the increased level of calcium in the β-cell triggers the release of insulin (insulin granule exocytosis) from the cell. The blood sugar concentration then decreases since insulin is a hormone that controls blood sugar levels by converting it to glycogen.

Diabetes arises from abnormalities in these release patterns. Diseases associated with mutations in K_{ATP} channels include unregulated insulin secretion and therefore low blood glucose levels. A mild form can be treated by the administration of a chocolate bar (favoured by children of all ages). This is an example (one of many) where Ca^{2+} ions act as a signal for a biological process.

Table 3.2 *Two second messenger ligands and receptors*

Ligand	Properties	Receptor
Ca^{2+} ion (a strange example to a coordination chemist of a metal ion acting as a so-called ligand).	Important second messenger. Certain K_{Ca} channels (BK, IK and SK) also open up to K^+ ions when bound to Ca^{2+} or calcium calmodulin. In the Ca^{2+}-activated K channel from *Methanobacterium thermoautotrophicum* a 'gating ring' made up of eight domains is attached to a channel similar to that in $K_{CS}A$ (3.4.2). Ca^{2+} ion binding to part of the ring results in widening out the entrance to the channel.	Ryanodine (Ry)(4). The receptor RyR is named after the plant alkaloid that can lock the receptor in the open state. It is a very effective Ca channel that is activated by an action potential and by Ca^{2+} binding to a specific site. It is thus called a Ca^{2+}-induced-Ca^{2+}-release (CICR) channel. The receptor exists in the SR of skeletal (RyR1) and cardiac (RyR2) muscle fibres and in other Ca^{2+} stores. Local anaesthetics (4.6) block RyR1.
Inositol-1,4,5-triphosphate IP_3	Important second messenger produced by phospholipase C catalysed splitting of the membrane lipid phosphotidylinositol-4,5-biphosphate PIP_2 into IP_3 and diacylglycerol (DAG), see Figure 3.6.	IP_3 receptors in EC coupling in muscle, in olfactory systems and involved in cell growth and division. These are IP_3 gated Ca channels. IP_3 binding triggers release of calcium. Resembles RyRs structurally, since it is also a CICR channel.

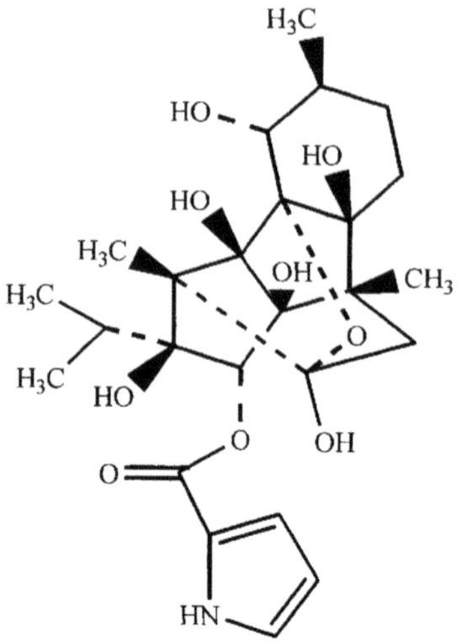

(4)

(4) *Ryanodine is extracted from the wood of Ryania speciosa. It is an insecticide that causes muscle contractions and acts by binding to calcium channels, promoting Ca²⁺ ion release and inhibition at low and high [Ca²⁺] respectively*

3.5.1 Ligands and Receptors

In contrast to voltage gated ion channels, ligand gated channels are usually not particularly ion discriminating. There are a variety of ligands, both extra- and intracellular that can open channels to ions (Tables 3.2 and 3.3). Two types of ligand gated channels are those that respond to second messengers and those that react to neurotransmitters.

Neurotransmitters are ligands released from the terminals of neurons that bind to specific receptors in the postsynaptic membrane of another neuron, muscle or gland cell. One of the most studied neurotransmitters is acetylcholine (ACh) which is the only one used at the neuromuscular junction (5.5), attaching to the specific receptor *n*AChR in the muscle cell. This receptor has been thoroughly studied and its structural aspects are shown in Figure 3.4. The general subunit topology and pentameric structure is observed for a number of mammalian ligand gated receptors *e.g.* GABA$_A$, glycine and 5-hydroxytryptamine (serotonin) receptors. Comparison of the activated (opened by ligand binding) state with the closed state, using cryoelectron microscopy, suggests that the α-helical segments making up the lining of the pore twist to effect the gating.

There is also a receptor for ACh, *m*AChR, present in smooth and cardiac muscles, with a muscarine (9) rather than a nicotine (10) agonist.

Cyclic nucleotide gated ion channels, again not particularly discriminat-

Table 3.3 *Some important neurotransmitters and their receptors*

Neurotransmitter	Properties	Receptor
Acetylcholine (ACh) (5)	Widely distributed in central and peripheral nervous systems. Released at nerve terminals and very important at the neuromuscular junction. Implicated in learning and memory.	nAChR (nicotine agonist, promoting channel opening, Na^+(in); K^+(out)). Skeletal muscle. Curare is a competitive antagonist. mAChR (muscarine binding). Cardiac and smooth muscles. No ion channel incorporated. Atropine is an antagonist.
γ-Aminobutyric acid (GABA) (6)	Inhibitory neurotransmitter in CNS, particularly the higher brain regions.	$GABA_A$ receptor, prevalent in brains of higher mammals. Anion (Cl^-) channel; GABA binding promotes rapid opening and hyperpolarization. Four different drug types bind (5.4).
L-Glutamic acid (7)	Most abundant excitatory neurotransmitter in mammalian brain. Involved with Ca^{2+}, in cell death.	Several types of ionotropic receptors permeable to Na^+, K^+ and Ca^{2+}. Metabotropic receptor for Ca^{2+} ions. First prokaryotic glutamate receptor was recently identified from the cyanobacterium *Synechocystis*. It has a K^+ selective pore.
5-Hydroxytryptamine (5-HT) (8)	Neurotransmitter and hormone in the CNS of invertebrates and vertebrates. Its precise role is uncertain, but it has been implicated in the regulation of sleep and awareness.	$5HT_3$ is a ligand gated Cl^- channel. Other 5HT receptors are G-protein coupled.
Glycine	Inhibitory neurotransmitter in the CNS, particularly the brain stem and spinal cord (amino acids are the most common neurotransmitters in the CNS).	Glycine receptor is an anion (Cl^-) channel. Strychnine (Table 5.5) blocks inhibitory action.

ing, occur in photoreceptors (7.4.2), olfactory sensory neurons (7.5.1), cardiac cells and other tissues. The tetrameric cGMP channel has four cGMP-binding sites and requires that all these sites be occupied for the channel to stay open almost all the time. Lower site occupancy leads to a much lower percentage of channel opening, *e.g.* triply occupied sites allow the channel to be open only 30% of the time.

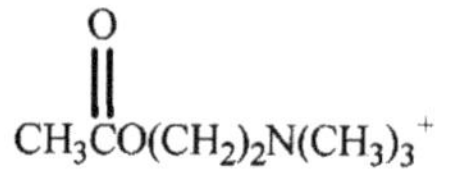

$$CH_3\overset{\displaystyle O}{\overset{\|}{C}}O(CH_2)_2N(CH_3)_3^+ \qquad\qquad H_2N(CH_2)_3CO_2H$$

(5)   (6)

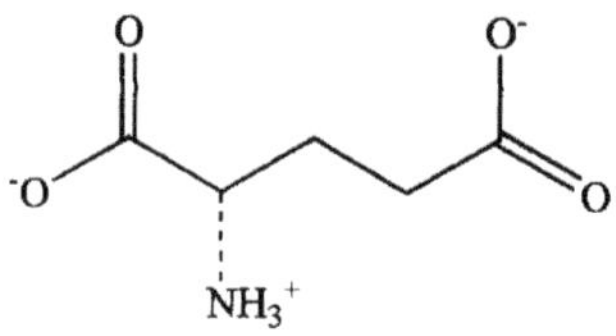

(7) 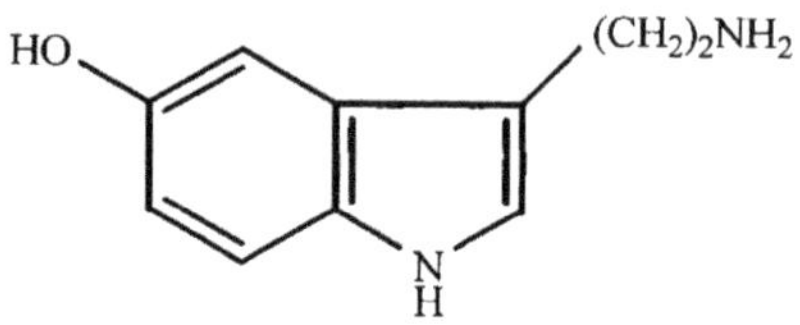   (8)

Most neurotransmitter receptors are *metabotropic*, that is, they do *not* incorporate an ion channel. In these, binding of a neurotransmitter will set off a chain of events that will lead to the eventual opening of an ion channel. A specific neurotransmitter can employ both an ionotropic and a metabotropic receptor for different purposes, *e.g.* GABA$_A$ and GABA$_B$ receptors respectively, are used by GABA. Glutamate binds and activates kainate (Table 5.3), *N*-methyl-D-aspartate, NMDA (11) and AMPA (12) receptors (ligand gated ion channels) or metabotropic glutamate receptors. ATP is an important neurotransmitter in the CNS and nerve–smooth muscle synapse, acting at ionotropic (nonselective for cations) and metabotropic purinergic receptors. These are distinct from the K$_{ATP}$ channels (3.5).

3.5.2 G-protein Linked Receptors

Many hydrophilic hormones, neurotransmitters, odorants and photons employ an indirect method for invoking an ion channel response. They use a guanyl nucleotide (guanosine triphosphate (GTP)) binding protein (called a G-protein) receptor that is specific for the stimulus. The whole entity can be represented as in Figure 3.5.

The sequence of events is as follows:

(a) G-protein bound to guanosine diphosphate (GDP), represented as G-GDP, and isolated from the receptor is in the resting ('off') state.

A.

B.

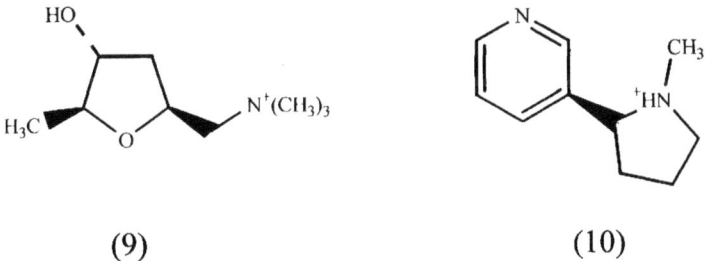

Figure 3.4 *(A)Topology of the nACh receptor in the nerve-muscle synapse. An extensive N-terminus in the extracellular domain binds agonists (e.g. ACh and α-bungarotoxin). There are four transmembrane domains with a long link between domains 3 and 4. This link contains the phosphorylation sites. There is a short C-terminus. The domain 2 helices from each of five subunits, $\alpha_2\beta\gamma\delta$, all of which contain four membrane spanning segments, form the pore lining for nonselective cations (Na^+ and K^+ ions). The channel opens upon ACh binding to both α-subunits. (B) Structure of the nACh receptor. Two (αs) of five subunits forming the funnel shaped structure. The remaining β (in front) and γ and δ (behind) subunits complete the receptor. The outer part of the pore protrudes some distance (~60 Å) from the membrane surface, while the inner part is only ~15 Å on the cytoplasmic side. The narrowest region is ~7 Å in diameter and only 3–6 Å long*

(9)   (10)

(9)(10) *L (+) muscarine (9) is shown. It is naturally occurring and is about 10^3 times more effective as a cholinergic neurotransmitter than is (−) muscarine*

(b) Agonist (mediator) binds to receptor R.

(c) R undergoes a conformational change and can now bind the G-GDP entity on the cytosol side.

(d) GDP is released from the G-protein in exchange for GTP at the α-site (now G-GTP). This is the entity shown in Figure 3.5.

(e) G-GTP breaks up into the active α-G-GTP complex and the β,γ dimer.

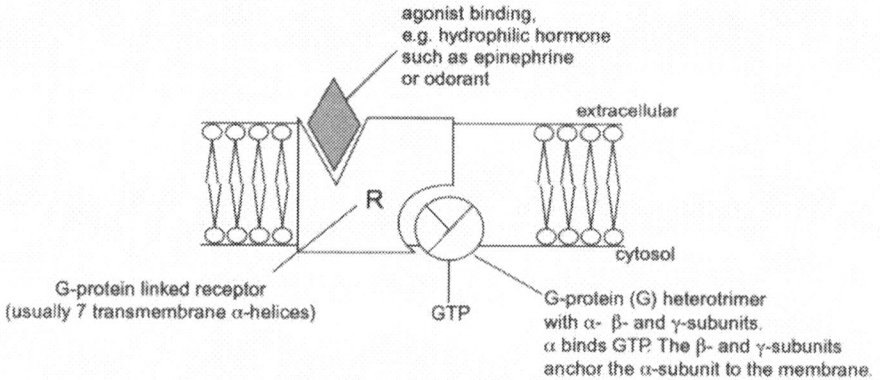

(11)   (12)

(11)(12) NMDA and AMPA are agonists (both simulate glutamate (7)) for subtypes of glutamate receptors. Activation results in the opening of associated ion channels and the generation of excitatory postsynaptic potentials (5.5). A strong stimulus is required for activation of the NMDA receptor in order to overcome blockage by Mg^{2+} ion

Figure 3.5 *Components of a G-protein linked receptor*

(f) To complete the cycle, the released α-subunit enzymatically hydrolyses the bound GTP to GDP and the original resting state (a) is resumed.

Before (f) can occur extensively the active α-G-GTP complex can trigger a second messenger (the agonist, *e.g.* hormone is the first messenger) in a variety of ways. Two of most interest to us are:

- α-G-GTP or β,γ dimer binds to the cytosol of an ion channel that opens. This is uncommon, but an important example is the G-protein gated cardiac K channel (6.4.5).
- A more complex, but more usual type, very pertinent to Ca^{2+} release, is shown in Figure 3.6.

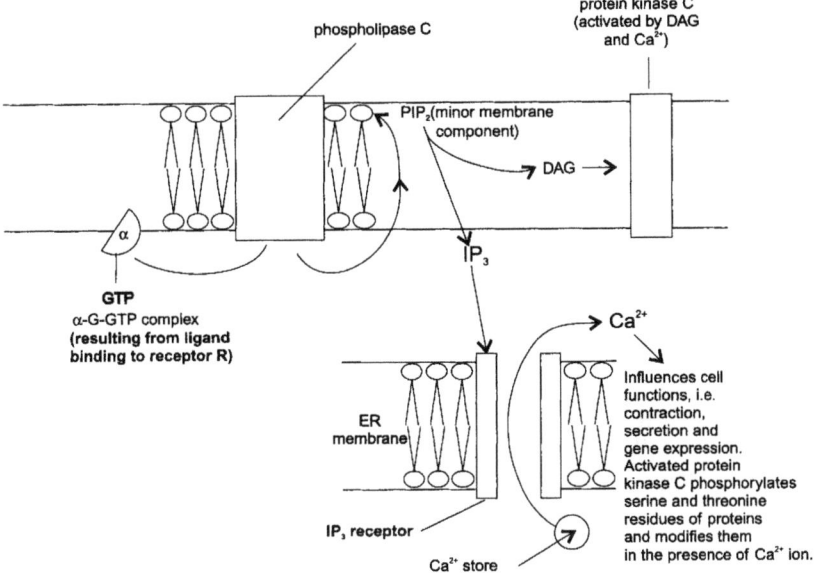

Figure 3.6 *The production of second messengers, IP_3, DAG and Ca^{2+} ions using signal transduction by G-proteins*

(13)   (14)   (15)

(13)(14)(15) *R_2 is often $CH_3(CH_2)_{16}$- and R_1 is often $CH_3(CH_2)_4(CH=CHCH_2)_4(CH_2)_2$-*

In this path, the active α-G-GTP complex activates phospholipase C (PLCδ1). All eukaryotic phospholipase C isozymes require Ca^{2+} ion for activity and the Ca^{2+} ion is localized at the active site. This enzyme catalyses the hydrolysis of the membrane lipid phosphatidylinositol-4, 5-biphosphate (PIP_2) (13) into constituents. The fragment inositol-1,4,5-triphosphate (IP_3) (14) enters the cytoplasm, moves to the ER and binds to an IP_3 receptor thereby opening the channel and stimulating the release of Ca^{2+} from stores (calcisomes). The synchronous production of IP_3 and Ca^{2+} ions has been demonstrated using fluorescent indicators (1.5.3). The other fragment of the hydrolysed membrane lipid, 1,2-diacylglycerol

(DAG) (15), remains in the membrane and like Ca^{2+} ions can activate protein kinase C. G-protein linked receptors have vital physiological functions and mutations of them are linked to a number of hereditary diseases or conditions, *e.g.* colour blindness. They are prime targets of about one quarter of the top selling drugs in the United States. Examples of these drugs are loratadine (Claritin) (16), an antihistamine antagonist of the H_1 receptor site, cimetidine (Tagamet) (17), an antiulcer drug, antagonist at the H_2 receptor site for turning off the production of stomach acid and doxazosin (Cardura) (18), an antagonist at the α_1-adrenergic receptors for reducing blood pressure.

(16)   (17)

(18)

Cholera Toxin Induces Potentially Lethal Dehydration

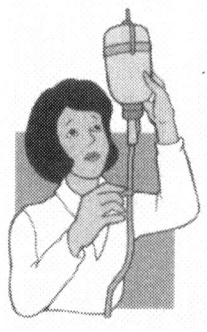

Cholera bacteria produce a toxin that inhibits the α-G-protein hydrolysis of GTP (step f). Cells therefore continue to produce cAMP from ATP (this is yet another second messenger produced from G-protein activation of adenylate cyclase (Table 7.2). Cyclic AMP stimulates active transport of Na^+ ions into the gut where the bacteria attack. The massive increase in Na^+ ion concentration and accompanying H_2O loss by osmosis leads to severe and rapid dehydration, ion imbalance and death within days without treatment (see Figure 2.1). See also 6.4.5.

3.5.3 Mediated Transport Systems

Specific transporters exist for the transport of organic compounds (X) across cell membranes (Figure 2.1). A change in the conformation of the transporter occurs continuously. As a result of the conformational changes binding sites and bound X are exposed first to one side of the membrane and then to the other. This affords a mechanism for X transfer through a membrane, the direction of which depends on the relative concentrations of X on either side of the membrane.

These systems are used for the transport of amino acids and glucose (see Figure 2.1). The transport is slow, only about 10^3 molecules of glucose per second across the membrane, for each glucose transport protein.

3.6 PUMPS

So far we have considered channels that use ion imbalance, *i.e.* that run 'downhill' along a charge or concentration gradient. Although the ions need help in crossing the membrane, no *direct* input of energy is required to effect the transport. However, when ions or polar molecules have to be moved from a low to a high concentration across a membrane, it is necessary to inject energy for successful channel operation. The system must run 'uphill' against an electrochemical gradient. This type of transfer is obviously necessary in order to maintain the ion imbalance across the membrane. Such channels are called *pumps* and they mediate *active transport*. There are two types of active transport, primary and secondary.

3.6.1 Primary Active Transport (ATPases)

ATP is an important source of energy in biological systems and the pump associated with it acts as an enzyme (ATPase) to catalyse the hydrolysis of ATP (ATP $\rightarrow$ ADP + P_i). In the process an aspartate residue on the pump is phosphorylated causing a conformational change that turns it inside out, allowing the transfer of attached ions from one side of the membrane to the other. The pump thus acts as a channel. It should be noted that all ATPase effects are modulated by Mg^{2+} ions. ATPases are involved in cellular processes such as transmembrane signalling, neuron impulse transmission, excitation–contraction coupling and secretion. These ATPase pumps belong to a superfamily of ion-translocating pumps in eukaryotic cells called P-types.

There are only four ATPases known to be linked to ion translocation. The directional flow (in parentheses) is that from the cell across the plasma membrane.

- *Na⁺(out)/K⁺(in) ATPase.* Probably the best known and most important of the ATPase pumps is the Na^+/K^+ ATPase pump, present in the plasma membranes of almost all animal cells where it controls ion balance. The structural features of the Na^+/K^+ ATPase pump are shown in Figure 3.7. Upon ATP binding, the protein pump undergoes a conformational change, which allows three Na^+ ions to be pumped out of the cell, while two K^+ ions are pumped in for each ATP molecule hydrolysed. In this case *both* ions are being moved against a concentration gradient. The necessary energy is provided by ATP hydrolysis. There is no transport unless ATP is hydrolysed and conversely there is no ATP hydrolysis without Na^+/K^+ transport. The

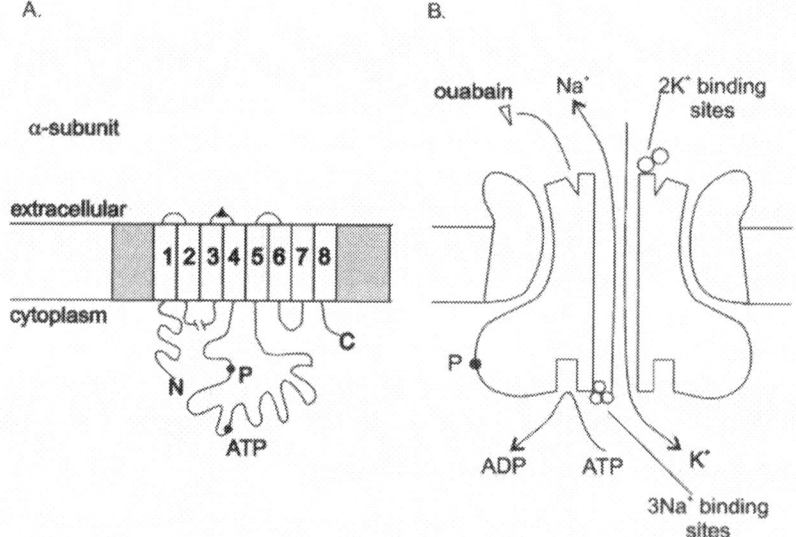

Figure 3.7 *(A) Topology of a Na^+/K^+ ATPase. There are eight transmembrane domains. The 4–5 cytosolic domain provides the PO_4^{3-} (P) binding site and the ATP hydrolysis site (ATP). The outer faces of domains 3–4 is the site (>) of cardiac glycoside (ouabain, digitalin) binding. These drugs do not inhibit other ATPases. (B) Structure of a Na^+/K^+ ATPase. Na^+/K^+ ATPase is a glycoprotein comprising four subunits ($α_2β_2$). The catalytic subunit (α) is in the cytosol and contains a site for dephosphorylation of ATP and another (P) for attachment of ATP γ-phosphate. The β-subunit is smaller and targets the pump to the plasma membrane*

Na$^+$/K$^+$ ATPase can drive non-voltage gated amiloride-sensitive sodium channels (Figure 2.1). Amiloride (19), even at nM concentrations, blocks an Na channel that is non-voltage gated and that is very Na$^+$ ion selective. In this way amiloride therefore acts as a diuretic and is also used for the treatment of hypertension. It is involved in taste modality (7.5.3) and is responsible for the low Na$^+$ ion concentration of the endolymph in the ear cochlea (7.3). In humans, mutations in this sodium channel can result in enhanced channel activity (hypertension) or reduced channel activity (hypotension). The channel has also been implicated in cystic fibrosis (2.1.2).

(19)   (20)

- *Ca^{2+}(out) ATPase* cytoplasmic pump (a high resolution X-ray crystal structure of which is now available) maintains a low concentration of Ca^{2+} ions in resting muscle, moving Ca^{2+} against a 10^4 concentration gradient using ATP hydrolysis (one ATP is dephosphorylated per two Ca^{2+} ions transported). It responds directly to Ca^{2+} ions in the cytosol and is important in skeletal muscle (6.2.1) and myocardiac contraction and relaxation (6.4.1).
- *H$^+$(out)/K$^+$(in) ATPase*, which is not as widespread as the Na$^+$/K$^+$ ATPase, is the other pump responsible for active transport of potassium ion across the vertebrate cell membrane. It promotes the supply of HCl by parietal cells in the gastric gland to the lumen of the stomach to aid in the digestive process. The chloride ion is provided by a HCO$_3^-$ (out)/Cl$^-$(in) pump (Figure 3.8). The parietal cells contain numerous mitochondria to generate ATP for powering the ATPase pumps. Omeprazole (Losec) inhibits the H$^+$/K$^+$ ATPase and is one of the world's biggest selling pharmaceuticals ever. It is used in the treatment of duodenal ulcers and reflux oesophagitis (acidic gastric juices that cause heartburn) by reducing excess stomach acid. The S-isomer, shown in (20) (esomeprazole), is several

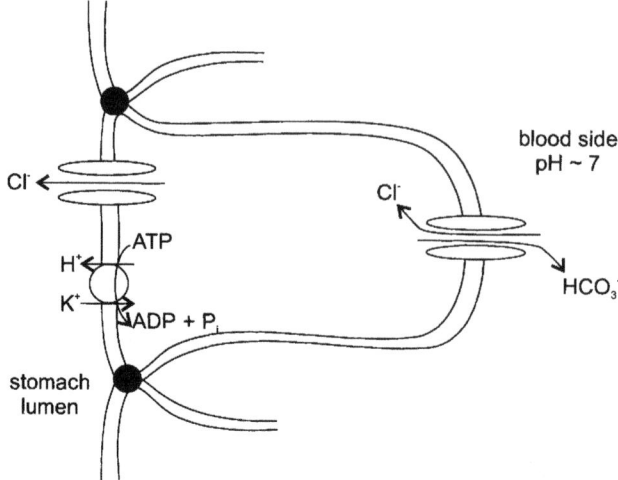

Figure 3.8 *Secretion of hydrochloric acid by parietal cells. In the cell $CO_2 + H_2O \rightleftharpoons H^+ + HCO_3^-$ is catalysed by carbonic anhydrase, which is present in high concentration in parietal cells. The Cl^- ions from the blood pass to the stomach lumen via a Cl^-/HCO_3^- antiporter and a Cl^- channel and combine with protons secreted into the stomach lumen in exchange for K^+ ions by a $H^+/K^+ATPase$. The combination of H^+ and Cl^- gives ~0.1 M HCl in the lumen. The hydrochloric acid accomplishes a number of tasks, i.e. the destruction of unwanted organisms, the denaturing of proteins so that they are easier to hydrolyse enzymatically and activation of these enzymes (pepsins from pepsinogens)*

times more potent than the *R*-isomer in humans and has recently been marketed as Nexium.

- $H^+(out)ATPase$. This is also called the proton pump. It is found in the plasma membrane and the membranes of several organelles. It is involved in the regulation of pH_I (Figure 3.9).

3.6.2 Secondary Active Transport

A source of energy other than ATP is one that derives from the 'downhill' concentration gradient of one ion moving another ion or solute against its concentration gradient. Thus, a 'downhill' transfer aids an 'uphill' one. The pair of species may move in opposite directions across the membrane (antiport) or move in concert (symport). The Na^+/H^+ and Na^+/Cl^- exchanges are examples of antiport exchanges. The $Na^+/glucose$ drip (2.1.2) and the Na^+ ion aided reentry of neurotransmitters to synaptic endings (5.3.1) illustrate symport exchange.

The Na^+/H^+ exchanger (NHE) is a transmembrane glycophosphoprotein, ubiquitous in mammals and widespread throughout the animal kingdom. It is an antiport, exchanging one Na^+ (in) for one H^+ (out), and is driven by the Na^+ electrochemical gradient. There are multiple isoforms, the NHE1 being the first to be cloned from human genomic DNA. Almost inactive when the intracellular pH ($pH_i = 7.1$) is at physiological value, it is turned on by several different stimuli. Neurotransmitters, peptide hormones and growth factors act within minutes, but acidic and osmolarity challenges are much slower (hours–days). The NHE aids the transfer of the antimalarial drug chloroquine into the cell of the malaria parasite *Plasmodium*. The NHE has a central role in cell volume regulation and in the control of pH.

The Na^+/Ca^{2+} exchanger is a *reversible* ion transport protein that is located on the plasma membrane of nearly all animal cell types. It is, for example, present in vertebrate photoreceptors and certain neurons, when K^+ is cotransported with Ca^{2+} (*i.e.* $4Na^+/1Ca^{2+},1K^+$). In most other cells $3Na^+$ ions are exchanged for $1Ca^{2+}$ ion, driven by the electrochemical gradient for Na^+ ($3Na^+$ in, $1Ca^{2+}$ out). In this condition Ca^{2+} is extruded from the cell, an occurrence that is used to maintain low $[Ca^{2+}]$ in cardiac myocytes. Occasionally more than two ions can participate. In a $Na^+/K^+/Cl^-$ cotransporter, Na^+ and Cl^- ions flowing inwards also drives K^+ ions inwards.

3.7 ION HOMEOSTASIS

The value of the many ion channels that we have encountered can be no better illustrated than by their capacity to maintain a relatively constant cellular environment. Two important examples are cellular pH and Ca^{2+} ion homeostasis.

3.7.1 pH_i Regulation

Strict control of pH_i is important since relatively small pH changes are likely to affect a number of processes including protein and enzyme activity in cellular metabolism, gene transcription, contractile strengths and others (Figure 3.9).

3.7.2 Cellular Ca^{2+} Ion Homeostasis

The maintenance of intracellular and extracellular calcium ion concentration in a resting cell is particularly difficult because of the large differences in these concentrations. Complicating the problem is the fact that the majority of the calcium in the cytosol and S(E)R is bound to

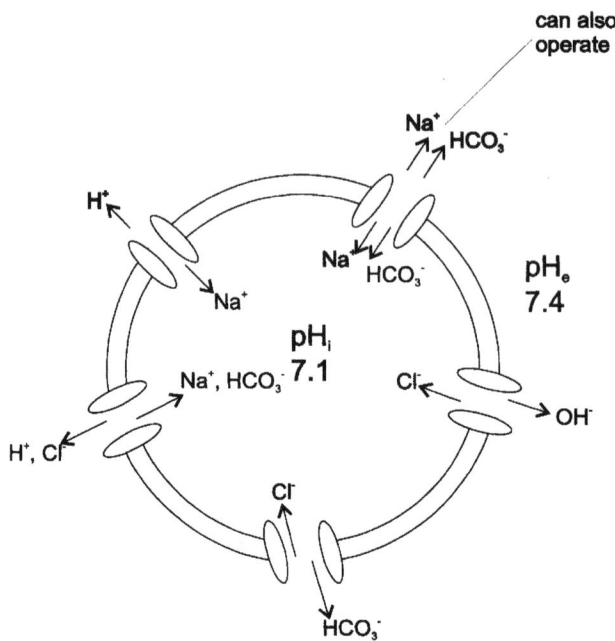

Figure 3.9 *pH$_i$ regulation by many ion channels. The regulation of pH$_i$ by ion channels in a variety of mammalian cells including cardiac myocytes, vascular smooth muscle, neuron and fibroblasts is represented. The constancy of pH$_i$ is remarkable and is regulated by proton pumps (H$^+$-ATPases) and proton channels. In addition there are several ion transporters shown in the figure. They include an Na$^+$/H$^+$ antiporter, an Na$^+$/HCO$_3^-$ cotransporter that mediates influx and efflux of HCO$_3^-$ in the cell and a Cl$^-$/OH$^-$ and Cl$^-$/HCO$_3^-$ exchanger. In addition to these in the myocyte membrane, in the kidneys there is a Na$^+$-dependent HCO$_3^-$ transporter that extrudes Cl$^-$ and H$^+$ ions in exchange for Na$^+$ and HCO$_3^-$ ions*

anionic sites on proteins, ATP, HPO$_4^-$ and/or citrate. Any changes in [Ca^{2+}]$_i$, which can rise 10-fold during its action as a second messenger, must be kept to brief bursts because of the toxicity of calcium ion.

The external calcium concentration is maintained *via* regulation by a complex hormonal feedback system. The internal calcium ion concentration results from internal and external feeds, the relative importance of which depends on the specific cell. All types of channels contribute to calcium ion homeostasis as shown in Figure 3.10. The Ca^{2+}/ATPase pumps maintain a low [Ca^{2+}]$_i$. These may be assisted by a Ca^{2+}/Na$^+$ exchanger, although the Ca^{2+} ion efflux can be turned into an influx depending on the Na$^+$ transmembrane electrochemical gradient. A Na$^+$/K$^+$ATPase controls this, in turn. Sodium and potassium ions therefore contribute to Ca^{2+} ion

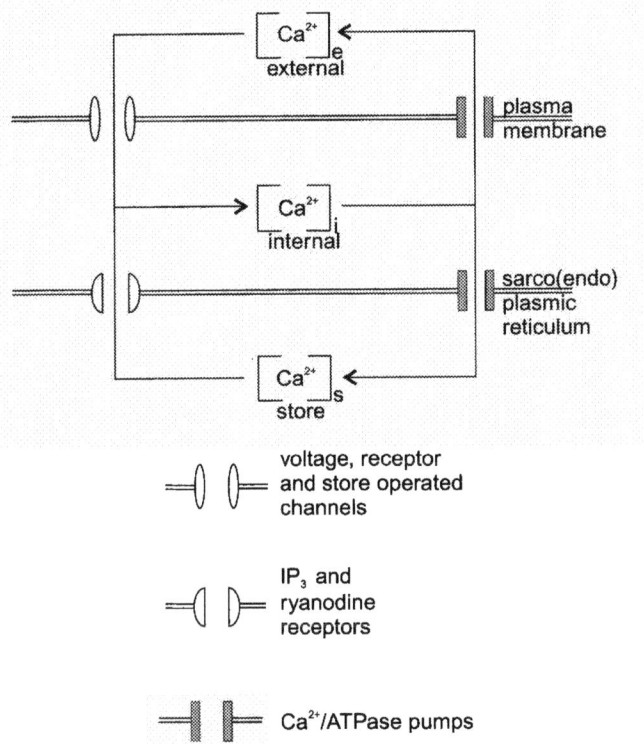

Figure 3.10 *Channels, receptors and pumps involved in Ca²⁺ homeostasis. Calcium is stored in mitochondria, the ER and the SR in striated muscles and in calciosome organelles in nonmuscle cells*

homeostasis. Internal stores feed Ca^{2+} ions into intracellular space by using receptors. These demonstrate Ca^{2+} ion-induced calcium release (CICR), but can convert to Ca^{2+} ion-inhibited calcium release if the increase in $[Ca^{2+}]_i$ could provoke toxicity. The delicate balance is upset when external conditions become extreme.

Extreme Hypothermia is not Necessarily a Killer!

Dropping the normal body temperature of 37 °C to around 23 °C stops both the heart and respiration, but the brain remains alive. Calcium ion levels rise on cooling because the Ca^{2+} ion pumps, which maintain low cellular $[Ca^{2+}]$, require energy and cannot function effectively in the cold. Rats and rabbits whose brains were cooled to 16 °C, were injected with EDTA, which reduced the Ca^{2+} ion level in the blood and hence in other cells. The animals started to function again and this continued on warming. Normally, warming is tricky since brain cells can warm up before blood starts flowing. Thus, cells can die from lack of O_2 and the animals would not survive. *If* this procedure were to work with humans, reviving before warming could be possible.

3.8 MOVING IONS BETWEEN CELLS

A rather different type of channel (*gap junction*) allows cations, anions and small molecules ($<10^3$ Da) to pass *between* the cytoplasms of adjoining cells in a variety of tissues. The plasma membranes of the two neighbouring cells are separated by a very narrow gap (about 2 nm) and contain proteins that are aligned opposite to each other to form an aqueous channel (Figure 3.11). The alignments are tight. Fluorescent dyes injected into one cell move into neighbouring cells without spilling over into the intercellular space.

Gap junctions between two neurons allow the very rapid flow of electric current bidirectionally using action potentials in the same way as within a cell. This property is used in certain large nerve cells (Mauthner cells) in fish brains, to effect rapid escape by using a tail flip. Electrical activity is

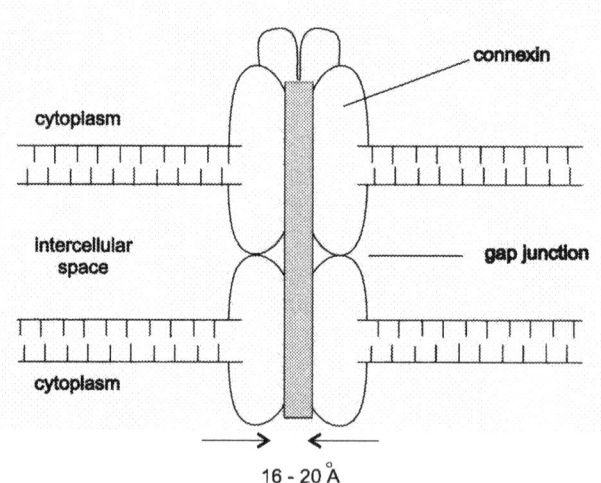

16 - 20 Å

Figure 3.11 *Gap junctions are constructed of transmembrane proteins (connexins). Six connexins form a cylinder with an open pore in the centre. These align with six connexins from the adjacent cell of the same type creating an open channel between the two cytoplasms that is usually considered to be a uniform tube. A connexin probably has four transmembrane domains S1 to S4. The two extracellular loops help tie the two hemichannels together*

conducted from fibre to fibre throughout certain types of smooth muscles (intestinal, uterine, small diameter blood vessels) by using gap junctions. These are particularly important in epithelial tissue and cardiac muscle. Gap junctions allow waves of contraction to spread through the muscle in concert (6.4.2).

In nonexcitable cells, gap junctions permit exchange of nutrients and act as a conduit for second messengers (*e.g.* cAMP and Ca^{2+} ions) between cells. Calcium ion influx into neurons can cause cell death. However, high cytoplasmic $[Ca^{2+}]$ can counter the influx by closing gap junctions between damaged and healthy cells within seconds and this slows further cell damage.

Reduced Channel Size Spells Disaster

Mutations in genes encoding gap junction proteins (connexions) – there are many mutants – cause a number of diseases including congenital deafness and Charcot–Marie–Tooth (CMT) disease. There are two forms of CMT, one arising from demyelination (4.5.1) and the other is a disease of the gap junction channel, varying from mild to severe (wheelchair requiring). For example, a S26L mutation in the connexin in the first domain S1 (which contributes to the channel pore) is believed to reduce the radius of the mutant channel to less than 3 Å (normally around 7 Å). This means that the mutant channel is probably impervious to cAMP and Ca^{2+} ions and so the single-channel conductance and effectiveness are reduced.

3.9 SUMMARY

Channels are the cell's windows to the outside world. They exist solely to allow the transport of small molecules. The movement of ions between the exterior and the interior of cells is at the heart of many processes that are required to maintain life. This means that a basically hydrophobic membrane, which encloses all cells, must be rendered pervious to charged species (ions). This is a difficult feat and membranes use a wide variety of channels and mechanisms for transferring the ions. Voltage changes, chemical agents and pumps are used to activate the proteins embedded in membranes. This activation leads to channel opening, either directly or indirectly, allowing the passage (sometimes specifically) of Na^+, K^+, Ca^{2+}, H^+ and Cl^- ions through the membrane.

CHAPTER 4

Intracellular Signalling

4.1 REFLEX ARC

Ions and ion channels play major roles in the conveyance of information from one part of an organism to another. In a vastly oversimplified picture, a signal would originate from the sense organ and be transmitted to the effector organ. The complete system is called the *reflex arc*. A schematic and a more specific example are shown in Figures 4.1 and 4.2 respectively.

Stimuli, both external (for example a cut or a noise) and internal (blood vessel distension) activate a sense organ. A network of nerve cells (neurons) links this to the effector organ (a muscle or gland cell). The network comprises:

- *Afferent neurons*, which are information gathering and carry electric signals (impulses) from the sense organ non-stop to the central nervous system (CNS). These are therefore bipolar cells, with one process in the periphery of the cell and a second branch joining up with the CNS. Different initiating senses and their associated afferent

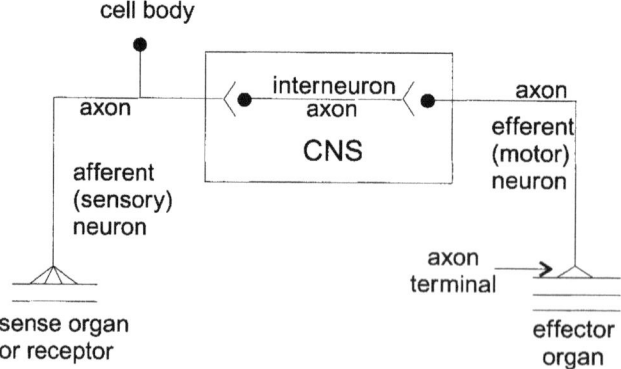

Figure 4.1 *Reflex arc*

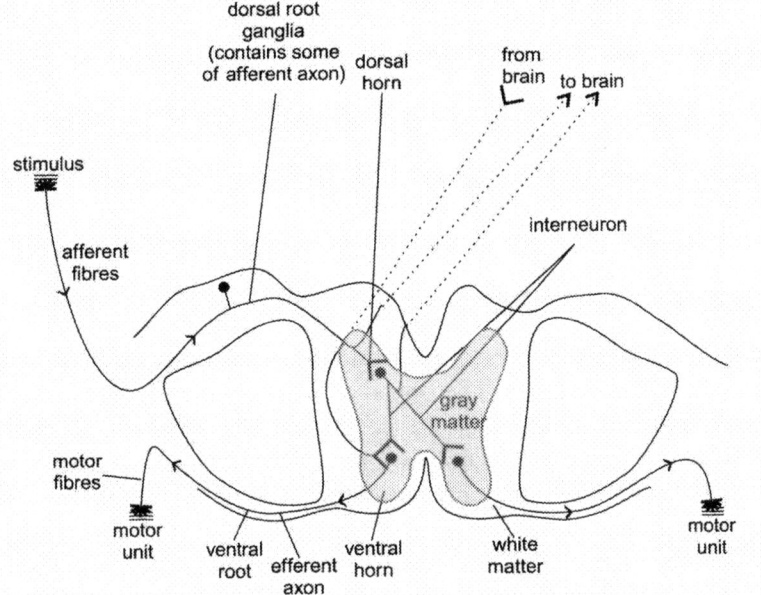

Figure 4.2 *A cross section through a single spinal segment. The spinal cord conducts signals between the brain, the stimulus and the effector organ (in this case muscle movement). Sensory fibres of afferent neurons enter the spinal cord in posterior or dorsal roots (back of body). Motor fibres leave the spinal cord via anterior or ventral roots (front surface of body). The fibres connect up via interneurons in grey matter, which contains cell bodies and dendrites of efferent (motor) neurons. Interneurons relay information to and from the brain, up and down the spinal cord*

neurons arrive at different places in the CNS. The CNS, consisting of the brain and spinal cord, processes the nature of the original stimulus.

- *Efferent or motor neurons* (sometimes called motoneurons) carry the impulses from the CNS to the effector organ.
- *Interneurons*, which always occur in the spinal cord and provide a link between sensory and motor neurons. Although interneurons are not always necessary they do comprise over 99% of the neurons in the human body.

The diagrams in Figures 4.1 and 4.2 are very simplified pictures. There are actually many interconnections at all junctions. The significant components are:

- Basic message transmission (electrical signalling) along the neurons.
- The transfer of information across the junctions.

• The connections between the sense organs and neurons and between the effector organs and neurons.

In this chapter we consider the electrical signalling along the neuron, which is similar for the three types of neurons in the reflex arc. Later chapters will deal with the other aspects.

4.2 NEURON STRUCTURE

There are a number of types of neurons that depend on the animal and their location in the creature. The one shown schematically in Figure 4.3 is the vertebrate spinal motor neuron (see Figure 4.2) and contains the main features of all neurons. There can be *many* dendrite and synaptic endings associated with an individual neuron when it is a multipolar neuron. A unipolar neuron has only one process arising from the cell body (see afferent neuron, Figure 4.1). Connections between neurons can be between soma or dendrites from either neuron. Many of the features in the figure are exaggerated.

There are four main regions in a neuron (see Figure 4.3):

(a) The dendritic tree consists of many dendrites and dendritic spines that arise from the cell body and are highly branched, A. They receive information as nerve impulses from neurons in other cells, which they relay to the cell body, B. Dendritic spines contain

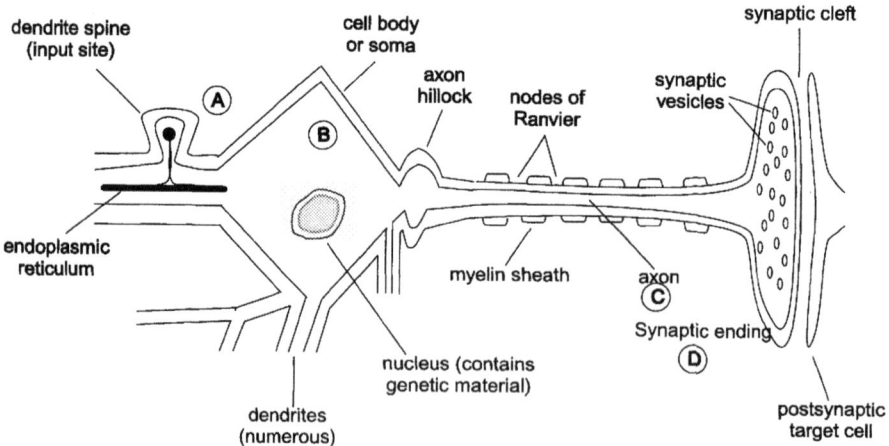

Figure 4.3 *Neuron structure*

receptor or voltage operated ion channels that are used to produce dendritic impulses.

(b) The cell body or soma contains the nucleus and the organelles that it needs to function (*e.g.* mitochondria, Golgi apparatus, *etc.*). It transfers the nerve impulses that it receives from a number of dendritic contacts down the axon, C.

(c) The axon is a single nerve fibre that functions as an electrical cable conducting impulses from the cell body, (set up at the axon hillock) unidirectionally, to the synaptic ending, D. The axon is usually thinner and longer (ranging from μm to metres) than dendrites. It has a myelin cover punctuated by breaks (nodes of Ranvier). A nerve is a bundle of nerve fibres.

(d) The synaptic ending transmits signals to target cells. These targets can be other nerve, muscle or gland cells. Synaptic transmission may be direct through gap junctions (3.8) or, as shown in the figure, by chemicals released from synaptic vesicles through synaptic clefts. Gap junctions are much narrower than synaptic clefts (which are 200–300 Å wide).

The endoplasmic reticulum (only a portion of which is shown in the figure) permeates the entire neuron.

4.3 NEURON CHARGE

First let us consider, in qualitative terms, a small isolated portion of the membrane enclosing the axon of a squid giant nerve. This was one of the earliest systems examined and is simple in that only K^+ and Na^+ ions and their respective channels need be considered.

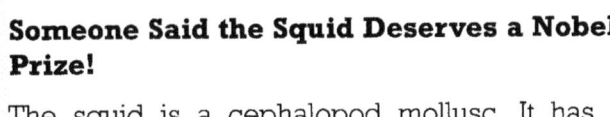

Someone Said the Squid Deserves a Nobel Prize!

The squid is a cephalopod mollusc. It has eight short tentacles around the mouth and two longer ones that can be used to shoot out and a seize prey. Specimens as long as 60 feet have been caught and one species, *Loligo architeuthis*, has been reported to grasp and sink a 150 ton schooner!

The *Loligo* squid has played a key part in the understanding of neuronal conduction and neuro-transmitter release at a chemical synapse. This is because it has a larger neuronal system than most vertebrates, containing *giant axons* and a *giant synapse* (Figure 4.4), into which electrodes can be inserted for direct measurements. A startled squid will activate its giant axons leading to muscle contraction in the mantle, emission of jets of water and rapid propulsion of the animal backwards – important both for attack and defence.

Recall the concentrations of K^+ and Na^+ ions inside and outside the axon (Table 1.1). If we first consider that only K channels (2P or 'leakage') are open in the membrane, then K^+ ions would pour from the inside of the axon across the membrane and would set up a positive charge on the outside of the axon membrane. Of course, a corresponding negative charge would result on the inside, with a consequent charge distribution across the axon membrane as shown in Figure 4.5.

The movement of K^+ ions would continue until electrostatic repulsion from the outside (or attraction from the inside) dominated, the net K^+ ion concentration change would stop, and a dynamic equilibrium would be established. Using the Nernst equation (1) (E in mV):

giant synapse
0.7 mm long

The stellate nerve
containing the giant
axon is several cm
in length and up
to 1 mm in diameter.
It connects to
the mantle muscle.

tentacles   siphon
              brain neurons
              integrating
              sensory modulaties

Figure 4.4 *Bare essentials of Loglio pealei – a common squid found along the North American coast*

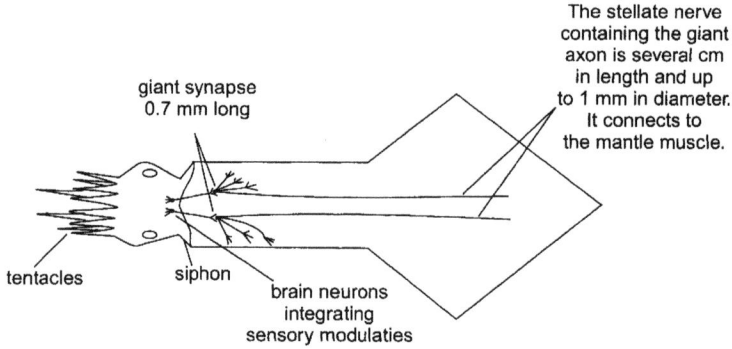

extracellular	low $[K^+]_o$	high $[Na^+]_o$	+ve
membrane			
	high $[K^+]_i$	low $[Na^+]_i$	-ve
intracellular			
	high $[K^+]_i$	low $[Na^+]_i$	-ve
membrane			
extracellular	low $[K^+]_o$	high $[Na^+]_o$	+ve

Figure 4.5 *Axon membrane charge distribution in a resting membrane considering only K^+ and Na^+ ions*

$$E = RT/F \ln[K^+]_o/[K^+]_i (= 58 \log[K^+]_o/[K^+]_i, \text{ at } 20\ ^\circ\text{C}) \qquad (1)$$

where the subscripts o and i denote the outside and inside concentrations, E, the inside voltage, can be estimated to be approximately -75 mV. This is the equilibrium potential for the system.

Now suppose we consider that an ion channel pervious only to Na^+ ions is operational. There would be a flux of Na^+ ions *into* the axon until, again, a dynamic equilibrium resulted. The inside would now be positive, and from the Nernst equation, the equilibrium potential would be approximately $+40$ mV. In a resting membrane, there will be a combination of contributions to the potential from all ions and channels present. Because the K channel is often open compared to the Na channel, the membrane is permeable mainly to K^+ ions (50–75 times more than to Na^+ ions), the major contribution to the so-called resting potential, V_{rest}, is the concentration gradient of K^+ ions. Therefore its value should be close to -75 mV, and is in fact -60 mV in the squid giant axon, and is relatively easily measured. The resting potential obviously depends on ion concentrations and membrane permeability differences and ranges from -20 mV to -100 mV in different animals. Thus, *all* cells at rest have an internal negative charge on the thin layer coating the inner surface of the membrane and a positive charge on the outer layer. We need not be concerned with a contribution from Cl^- channels to the resting potential, since the permeability of the membrane at rest to Cl^- ions is approximately zero; *i.e.* the channel is closed. However, chloride channels do make important contributions when the nerve is working. There will be an influx of Cl^- ions with open channels and E_{Cl} will equal -42 mV.

4.4 GENERATION OF ACTION POTENTIALS

Consider now the operation of two additional types of channels, one sensitive to Na^+ and the other to K^+ ions. These are voltage gated, that is the probability of their being open depends on the value of the membrane potential. Furthermore, the Na channel opens and closes in less than a millisecond and remains inactivated for some time until the membrane is hyperpolarized (see below). The K channel, called a voltage gated delayed rectifier, responds more slowly, over several milliseconds. Under an electrical stimulus these two channels are brought into action. This stimulus, a localized potential (also called synaptic, end-plate, pacemaker or receptor potential depending on its location) spreads over very short distances (nm or so). If the electrical signal is sufficiently strong (about a 15 mV reduction in electrical potential difference across the membrane, termed a *depolarization*) the Na channels will open. As the positive charge added to the cell increases, more and more voltage gated Na channels will open. This effect is shown in Figure 4.6, in which the patch-clamping technique is used (Figure 4.7).

The momentary flow of Na$^+$ into the inside of the axon counteracts the negative resting potential and the Na$^+$ equilibrium potential of +40 mV is almost reached before the Na channel shuts. This phase is called *depolarization*. Meanwhile, the slower opening of the K channels begins to dominate and the efflux of K$^+$ ions reverses the potential change (*repolarization*). In fact, the membrane potential briefly becomes more negative than the resting potential and almost reaches the K$^+$ equilibrium potential (*hyperpolarization*) before reversing and slowly regaining the resting potential in the *afterhyperpolarization* phase. The whole series of events is called the *action potential*, often called the 'spike' (Figure 4.8).

The operation of a threshold stimulus (localized potential) for promoting an action potential is necessary to ensure that *small* random changes in the

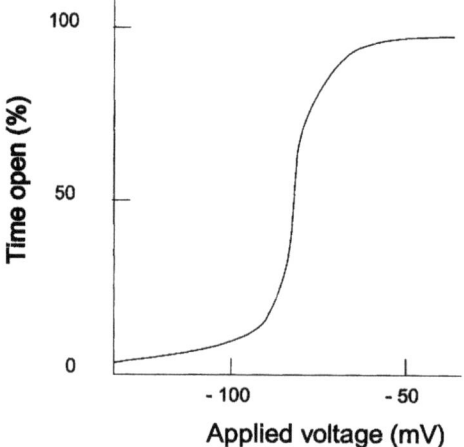

Figure 4.6 *Patch clamp experiment on an isolated Na channel. Channel open time is plotted as a function of applied membrane potential using the patch-clamping technique (Figure 4.7). Purified reconstituted rat brain Na channels were incorporated into an artificial membrane (planar lipid bilayers). Single channel currents were measured under voltage clamped conditions with 1 mM batracho-toxin added. This is a specific Na channel activator that eliminates inactivation by shifting the voltage dependence of activation in the hyperpolarizing direction (Table 4.4). As the applied voltage is made more and more positive, the channel opens more frequently (determined from the relative times current is flowing and stopping, i.e. on and off). It can be seen that with depolarization to −80 mV, the channel is open 50% of the time and at −60 mV it is almost always open. It was shown that the Na channel survives the transfer from in vivo to in vitro conditions unchanged*

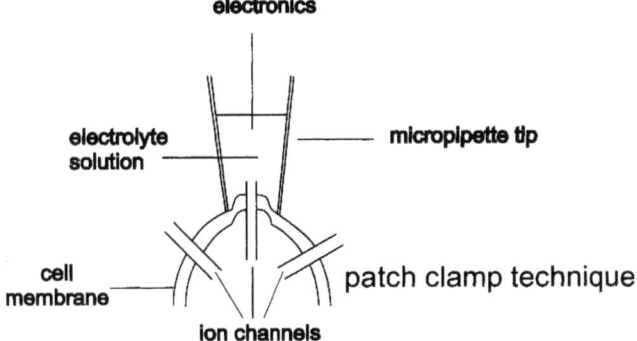

Figure 4.7 *Channel examination. In the figure, a patch of membrane is isolated by pressing the tip of a micropipette (tip diameter = 1–2 μM, less than 10⁻⁴ the diameter of a human hair) against a surface of a cell and applying gentle suction, thus giving a tight high resistance seal between pipette and cell surface that is pulled into the pipette. A single channel can even be isolated in this way. When the channel is open, current flows and is detected by a sensitive electronic circuit (current to voltage converter connected to the inside of the electrode). It is also possible to pull the pipette and membrane portion off the cell rapidly so that the isolated channel can be examined free from the cell. Whole cell patch clamping is a variant in which the patch membrane is destroyed (by strong suction). This allows access to the whole cell. The voltage clamp technique allows the measurement, at a set membrane potential, of the current following the opening of thousands of channels. This was vital in reaching an understanding of the action potentials of the squid and other axons. Originally impalement of the membrane with microelectrodes was used, but now voltage clamping is combined with whole cell patch clamping. This allows the examination of mammalian and other small cells that are still in contact with their neighbours*

membrane potential, which always occur, do not register as meaningful transmittable action potentials.

In the hyperpolarization and afterhyperpolarization phases other types of K channels, both voltage gated and Ca^{2+} ion activated (SK and IK channels), are brought into action. The total conductance change during an action potential can be separated into the individual components attributable to the opening of Na and K channels by using specific channel blockers (TTX for Na and TEA for K).

Plant Eaters Crave Help With Their Nerves

The low concentration of Na^+ ion in the haemolymphs of moths compared to that of K^+ (Table 4.1) could present a problem in the generation of an action potential for nerve signalling. The successful operation of an action potential requires a high

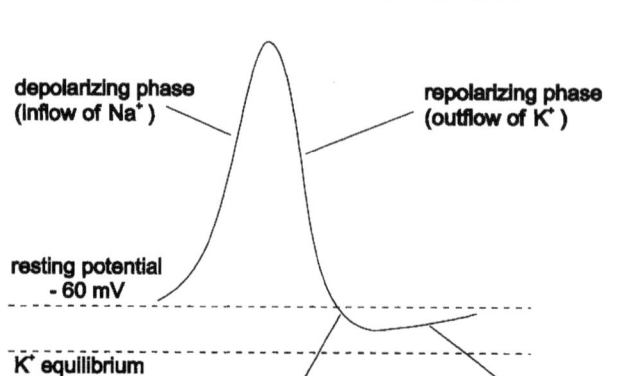

Figure 4.8 *Action potential. The change of membrane potential upon electrical excitation. The voltages (currents) are measured by inserting stimulating and recording electrodes into the axoplasm (axon cytoplasm). The total conductance change reflected in the action potential can be separated into contributions from voltage gated Na and K channels using specific channel blockers. Blocking Na channels with TTX isolates the (later) delayed rectifier K current change. Using the K channel blocker TEA allows the examination of only the early Na current*

Table 4.1 *Haemolymph concentrations of sodium and potassium ions*

Insect	[Na$^+$], mM	[K$^+$], mM
Cecropia moth	2.5	54.0
Cockroach	161.0	7.9

extraneural Na$^+$ ion concentration. This is achieved in most insects by using a nerve sheath (perineurium) that protects the nerve from direct contact with extracellular fluid so that the [Na$^+$] can remain high in the perineurium (*via* an Na pump). In contrast, the cockroach, which is omnivorous, has the desired high Na$^+$ concentration in its haemolymph.

It can be easily shown that the actual *amounts* of K$^+$ and Na$^+$ ions transferred across the membrane during an action potential are very small (~1 per 10^6) compared with the amounts present. There is thus little disturbance of the ion concentrations inside and outside the membrane. Eventually, an Na$^+$/K$^+$ ATPase can remove three Na$^+$ ions for every two

K$^+$ ions taken in, and reduce any substantial imbalance. A squid axon can generate thousands of impulses even if this pump is inoperative.

The simple behaviour described above for the squid giant axon is atypical. Examination of Table 1.1 shows that an influx of Ca^{2+} ions (like Na$^+$ ions) would occur if voltage gated Ca channels were operative ($E_{Ca} = +135$ mV). If these were activated and inactivated more slowly than Na channels, the depolarization phase would be much longer than that observed for the squid giant axon. This situation occurs in heart muscle cells where L-type voltage gated Ca channels control the influx of ions rather more than do Na channels (Figure 6.10). Ca^{2+} ion currents also carry the inward currents in vertebrate smooth muscle cells, certain crustacean (crab and crayfish) muscles and celiate (paramecium) nerve cells. These action potentials can be as long as 200–300 ms in duration compared to 1–5 ms in nerve fibres or skeletal muscle and 10–100 ms in invertebrate neurons. Although we have used the neuron to illustrate the generation of an action potential, skeletal and cardiac muscles and even single cell organisms (paramecia) can also produce action potentials.

Large Charge out of *Electrophorus*

The electric eel (freshwater) and the electric ray (salt water) are startling if unusual illustrations of the voltage changes developed in membranes *via* action potentials. The power of the electric eel was demonstrated as early as the eighteenth century when ten people holding hands received severe shocks when a moderate sized eel was included in the circle! The electric eel (*Electrophorus electricus*) lives in South American rivers, reaches 8–10 feet in length and is the most powerful bioelectric generator in the animal kingdom.

About half of the eel consists of electroplates that are stacked in many columns, each containing 5000–10 000 plates, and running parallel to the spinal column. The columns of plates are insulated from one another to prevent short-circuiting. The electroplates are modified muscles that cannot contract since only the posterior membranes have nerve endings and are excitable (Figure 4.9). Resting potentials (–70 mV to –85 mV) are maintained by K$^+$ ion channels (and Na$^+$/K$^+$ATPase) in both membranes. Stimulation (electrical or chemical) causes release of acetylcholine (ACh) onto the innervated membrane. End plate potentials produced by ACh receptors surpass threshold and trigger action potentials (Na channel generated) with peaks of ~ +70 mV. The noninervated membrane

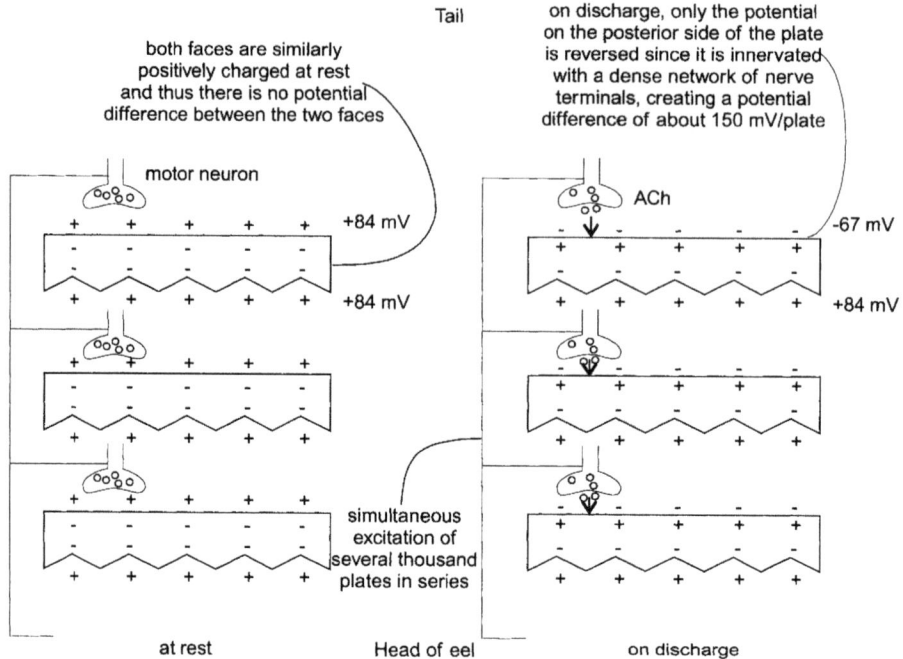

Figure 4.9 *Discharge of the electric organs of the electric eel*

contains no AChRs or voltage gated Na channels and maintains the resting potential of −84 mV. Thus, approximately a 150 mV difference between the faces of each plate is generated. The associated current flows from innervated to noninervated membrane, out of the head of the eel through the water and back into the tail region. An overall voltage discharge of (~0.15 × 4000) ~600 volts is generated that is used to stun prey. The creature can operate at frequencies as high as 400 pulses s⁻¹ for days! Similar electroplates in other parts of the eel are spaced some distance apart. These produce small, single non-stimulated releases and could be used for navigation.

Electrophorus produce large quantities of Na⁺/K⁺ATPase, have many Na channels and *n*ACh receptors. The eel then is an important source of the proteins for electrophysiological and structural exam-inations. The marine electric ray (genus *Torpedo*) works on the same principle. It generates a much lower voltage (about 80 volts), but higher amperage because of the conductivity of seawater. In addi-tion, it tires quickly finding 10–20 discharges exhausting! It has packed nAChR in its electroplate membranes and has historically been an important source of the receptor for study.

4.4.1 The Propagation of Action Potential Along an Axon

We now have to understand how the action potential can be used to transfer charge along the axon and thereby generate a (signalling) current. Only in certain cells, called electrically excitable cells (*e.g.* neurons, cardiac cells and skeletal muscle), can the membrane be used in this way. Nonexcitable cells, such as blood and endothelial cells do not excite action potentials but still posses all types of ion channels. In the resting state the distribution of charge along the axon is as represented as A in Figure 4.10.

A region of depolarization initiated at the cell body end shown in B results in the production of a potential of about +40 mV and a charged state. The charge causes an immediately joining patch to be partially depolarized as in C, the sodium channels open and the action potential cycle is repeated. The original fragment then returns to its resting state also shown in C. Continual repetition results in eventual charge transfer from the cell body near the axon hillock to the nerve terminal end. This *passive spread* allows the propagation of charge (*i.e.* electrical signal) over long distances, even metres, along the axon. The transmission of the action potential has been likened to that of a spark in an explosive fuse. The lighting of the fuse causes a small amount of gunpowder to explode. The neighbouring portion of the fuse heats up and the gunpowder there explodes, which in turn supplies the heat for the next section, eventually leading to a blast.

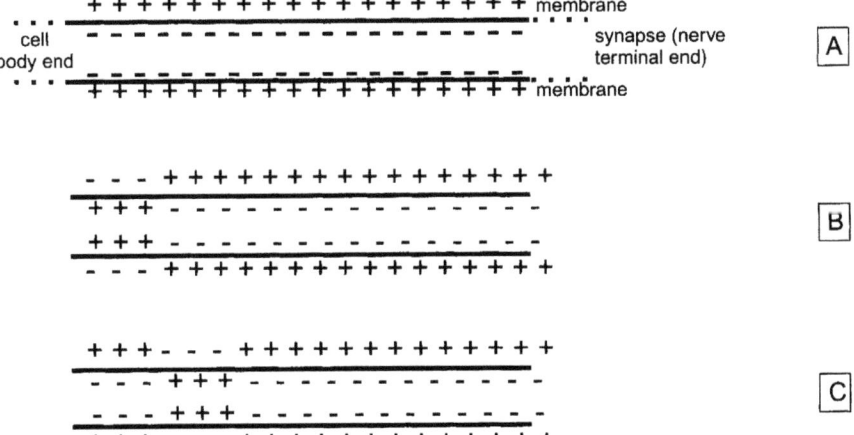

Figure 4.10 *Propagation of action potential along an axon*

Surprisingly few processes are involved in electrical signalling and in general these are very similar in such creatures as jellyfish, mice and humans! It is logical that the opening of Na channels will play a major role in the speeds of neuronal conduction to the brain and muscle fibres. There is a direct relation between the speed of an animal and the reaction time of its Na channels.

Squids Show Lightning Speed Compared With Their Relatives

Voltage gated Na channels in nerve cells are opened as they transmit current pulses racing along neurons to the brain and muscle fibres. Whole cell patch clamp recordings of Na currents from similar neurons in cephalopod (*Loligo*) (*e.g.* octopus, squid) and gastropod (*Aplysia*) (*e.g.* sea hare, snails) species are shown in Figure 4.11.

The sodium channel of the squid giant fibre opens ten times faster (and can repeat this 200 times s^{-1}) than that of the sea hare, which also has a slower repeat rate (30 times s^{-1}). This quicker electrical impulse transmittance to the muscle enables the squid to make a faster getaway even though it has an internal structure similar to that of the slug (and snail). The rapid getaway (10 km h^{-1}) is achieved by using water jets from the large mantle cavity (4.3). Also important to its speed is the presence of a wide submyelinic space and long

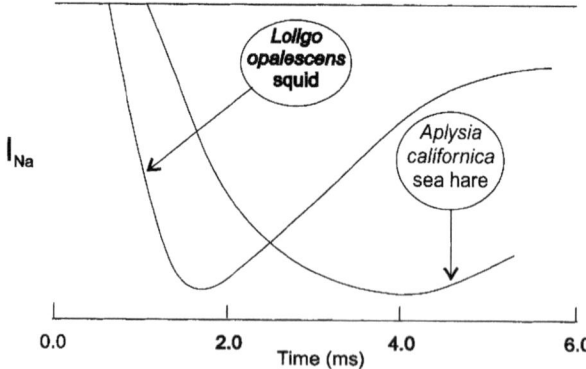

Figure 4.11 *Sodium currents in voltage gated Na channels in moluscan neurons. Whole cell patch clamp recordings with the voltage held at −10 mV are shown. The whole cell sodium current (I_{Na}) is the sum of currents passing through all the Na channels in the plasma membrane*

internodal distances that allow saltatory impulse conduction (4.4.3), which is rare in invertebrates.

4.4.2 Patterns of Action Potential Firing

There are two refractory periods following an action potential. In the *absolute refractory period*, there is total inactivation of the Na channels and continued activation of the K channels. An action potential is impossible under these conditions, regardless of the stimulus strength. Following this is a *relative refractory period* where depolarization in excess of normal is necessary to open the Na channels. This latter period, lasting about 10–15 ms or longer, coincides roughly with the hyperpolarization period and now action potentials can again be fired. This delaying period is important because it limits the firing frequency of repetitive action potentials (to ~100 or so per second) and is essential for normal neurotransmission. In addition the refractory periods ensure that charge movement is unidirectional from the axon hillock. As long as a stimulus is maintained, axons will fire action potentials, but they will be spaced in time, because of the refractory periods. When the stimulus (depolarizing voltage applied) is larger, the frequency of action potentials is greater.

Neurons do not only fire strings of action potentials at constant frequency. They also show complex patterns depending on the neuron function. *Aplysia* exhibits a good example of this in its inking behaviour.

Aplysia Muddies Up the Water

Aplysia is a species of marine mollusc, which is also called sea slug or sea hare (because of its appearance). It is very useful in neurobiological research because of the relatively small number of large, easily identifiable, neurons it possesses.

The motor neurons in the ink gland of the *Aplysia* mollusc are responsible for ejecting jets of purple fluid after strong provocation (*e.g.* puncture by a predator). The ink blackens the surrounding water and could allow the sea hare to escape. Healthy *Aplysia* never ink spontaneously. Obviously, the neuron response that produces the jets of ink is neither required, nor even desirable for small, intermittent signals. Neurons, therefore, fire only when the stimulus is large (a threshold of about 30 mV depolarization) and sustained and then only after a delay of some one to three seconds. In contrast, the gill motor cells (responsible for gill retraction) are spontaneously active and have low thresholds of about 5 mV depolarization.

4.4.3 Myelin Covering

Glia (or neuroglia) support neurons in the CNS. One of the four classes, oligodendrocytes, form sheaths of myelin around axons. These insulate the axon from other axons and aid action potential propagation. Sodium channels are sparse in the myelin membrane covering, whereas K channels are plentiful. At the axon hillock and the nodes of Ranvier voltage gated Na channels abound and therefore it is only here that action potentials can be initiated and generated. The action potentials 'jump' from node to node, which are spaced about 1–2 nm apart along the axon that is here exposed to extracellular fluid. This 'jumping' (called saltatory conduction) allows propagation of action potentials (conduction speeds) to be much faster in the myelinated axons of vertebrates compared with the nonmyelenated axons of vertebrates and invertebrates. There is a large variation of conduction (signalling) speeds among motor nerves of animals, for example (m s^{-1}): 0.8 (snail), 3–36 (fish) and 30–120 (cat). The largest myelinated axons in humans conduct impulses at rates as high as 120 m s^{-1} (270 miles per hour) (see also 7.7.1). Although vertebrates have small axons, and there is, in general, a direct relationship between velocity of conduction and axon diameter and myelination overrides this disadvantage.

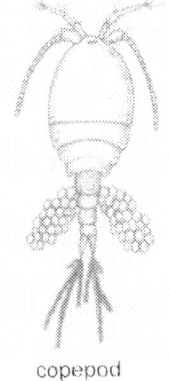

copepod

Crafty Copepods Can Avoid Potential Foes

Copepods are minute crustaceans that are a large part of marine zooplankton. Copepods of the order *Calanoida* have nerves with myelin-like sheaths, and these allow rapid signalling and fast evasive action for the creature. They can accelerate up to 200 body lengths s^{-1} within ms! This, in part, has enabled them to outrun foes and become the most abundant metazoan in the oceans. Myelinated axons in invertebrates are rare, but occur in squid (above) and also in penaeid shrimp, which exhibit conduction speeds of 200 m s^{-1}, among the highest in the animal kingdom.

4.5 MALFUNCTIONING OF NEURONS

All parts of the neuron are liable to malfunction. This is likely to result in clinical disorders, which are often severe, arising from the disturbance of neuronal signals in the nerves or brain and often also in abnormalities in muscle behaviour or even paralysis. The causes can be either internal or external.

4.5.1 Intracorporal Upset

The loss of an abnormal myelin sheath can lead to severe disorders in humans.

Disastrous Demyelination

Demyelinating diseases include Charcot–Marie–Tooth (CMT) (weakness of hand and foot muscles) and multiple sclerosis (loss of muscle control). CMT is an inherited peripheral neuropathy in about 1 in 2500 individuals. As well as resulting from mutations in genes encoding myelin proteins, one form is linked to mutations in genes encoding gap junction proteins (3.8). The slowing of axonal conduction also shows up in canine distemper where a virus kills glial cells.

Any mutation of an ion channel gene can cause neuron disorders. We have already briefly alluded to the strange behaviour of mice with mutations in genes encoding for each of the two subunit types (α and β) in neuronal voltage gated Ca channels (3.4.1). Examples of human diseases arising from abnormal K, Na and Ca channels are shown in Table 4.2. A number of human diseases have now been ascribed to such ion channelopathies.

Epilepsy Abounds in an Australian Family

Epilepsy is one of the oldest recognized human disorders, it having been mentioned in 3000-year-old Babylonian writings. Epilepsy is present in about 1% of the general population. It results from excess activity in brain neurons and the symptoms include loss of consciousness, sustained or rhythmic muscle contraction and hallucinations. The different forms of epilepsy can be linked to mutations in either Na or K voltage gated channels. There is a remarkable family in a remote area of Australia, descended from early English settlers, in which there has been inbreeding. In one sixty-member branch of the family, twenty-three individuals were epileptic, experiencing both non-fever and fever-induced seizures (generalized *epilepsy with*

Table 4.2 *Malfunctions in voltage gated neuronal channels*

Site	Channel Gene	Result	Disorder
K channel in brain and peripheral nerves.	KCNA1. This gene is the mammalian homologue of the fruit fly *Shaker*. Mutations in the first, fifth and sixth transmembrane helices in different families.	Delays repolarization, producing abnormal spontaneous action potentials.	Episodic ataxia with myokymia (EA-1) is a dominantly inherited disorder with limb uncoordination (min. to hr.) and muscle rippling. Provoked by sudden stress.
Na channel in brain.	SCN1B encoding β_1 auxiliary subunit.	Long, late openings of Na channels.	Generalized epilepsy with febrile seizures.
Brain specific Ca channel.	CACNL1A4 encoding α_1 subunit of P/Q type Ca channels.	Still uncertain, but altered ion selection likely.	Hemiplegic migrane (familial form of headache). This is one of three diseases arising from different types of mutations.

febrile seizures). The SCN1B gene of the epileptic family members was found to have a single base pair substitution at position 121 (cys replaced trp, C121W) in the β_1 auxiliary subunit of the voltage gated Na channel. The β_1 subunit normally increases the rate at which the channel opens and closes, supplementing the rate of inactivation of the Na channel principal α-subunit. The mutation destroys the β_1 function, slowing Na channel opening and closing, which results in the observed neuronal hyperexcitability. These types of findings form the basis of another approach to the treatment of epilepsy. The use of drugs to combat epilepsy has had limited success. Phenytoin (1) is the most widely prescribed antiseizure drug. It acts by blocking the persistent Na^+ ion currents. No mutations in α-subunits of neuronal sodium channels have as yet been found to cause an inherited disease.

(1)

4.5.2 Extracorporal Invasion (Neurotoxins)

All the neuron voltage gated channels considered so far, particularly the Na channel, can be targets of neurotoxins. The toxins can function by either blocking the channel in several ways or occasionally by keeping it open. In all cases, action potential transmission is markedly modified and the neuron system no longer functions correctly. This will show up as muscle failure since neuron and muscular function are intimately linked (5.5). Species such as scorpions, cone snails and funnel-web spiders usually employ, in one venom, a cocktail of neurotoxins that can interfere with the actions of a variety of channels. This acts as insurance against the different types of prey it may encounter.

A wide variety of species use neurotoxins in both attack and defence. They include primitive dinoflagellates, spiders, scorpions, frogs, bees, fish and snakes. Some plants also use neurotoxins as protection against insects. Many different compounds are used as neurotoxins. These include heterocyclics, alkaloids and, by far the most popular, polypeptides comprised of 17–64 amino acids. Many neurotoxins are commercially available for use in the characterization of ion channels, receptors and attendant enzymes.

Certain Turtles Can Withstand Prolonged Anoxia by Cutting Down on Ion Channels

Following anoxia, a cascade of degenerative events, including loss of electrical activity and ion gradient in mammalian brains, leads to death. In contrast, certain species of turtle seem better able to withstand prolonged anoxia lasting over weeks or even months. This is ascribed in part to a depression in the activity of key neuronal ion channels and a concomitant conservation of energy (>50%). It has been found, for example, that exposure of the cerebellum of the freshwater turtle *Pseudemys scripta elegans* to four hours of anoxia produced a 42% decrease in voltage gated Na channel density. This value was determined by measuring (radio-activity) the decrease in the number of [^{3}H] brevetoxin binding sites resulting from anoxia. Brevetoxin produced by marine dinoflagellates acts like batrachotoxins (Table 4.4) and has a high affinity binding to sodium channels.

A sample of the wide range of neurotoxins and the channels that they target is presented in Tables 4.3–4.6. Later, toxins that attack other parts of the nervous system will be described.

Captain Cook Was One of the Lucky Ones!

Many toxins single out sodium channels and target a number of different sites for attack. One of the most deadly neurotoxins is tetrodotoxin, TTX, (Structure (1), Chapter 3) which is unusual in not being a polypeptide. It is found in a wide range of creatures but is mainly associated with the Japanese puffer fish. Its origin is uncertain but could be promoted by symbiotic bacteria. Puffer fish raised in hatcheries do not contain TTX (but do they taste as good?). TTX, effective even in nM concentrations, blocks the voltage gated Na channel from the outside, favouring the pore region between S5 and S6 in the four domains (Figure 3.2). This prevents conduction of nerve impulses along the axon, leading to paralysis of respiratory muscles and usually causes death within hours (however, Captain James Cook survived TTX poisoning). Since TTX cannot cross the blood–brain barrier, the victim remains horribly conscious while his peripheral nervous system shuts down! The guanidinium moiety of TTX (Structure 1) binds to a glu residue at the channel mouth and forms an effective plug. It is probable that TTX cannot

Table 4.3 *External sodium channel blockers*

Toxin and Chemical	Source	Physiological Effect	Clinical Effect
Tetrodotoxin (TTX), heterocyclic	Main source is Japanese puffer fish.	Blocks nerve and muscle action potentials (has no effect on K channels).	Respiratory muscle paralysis.
Saxitoxin heterocyclic	Marine dinoflagellates (large groups are called 'red tide') Eaten by shellfish that then become poisonous.	Blocks nerve and muscle action potentials.	Respiratory muscle paralysis.
μ-Conotoxins 17–31 amino acid polypeptides	Predatory marine *Conus* snails. Injected into fish by disposable tooth. Beautiful shells!	Blocks mainly skeletal, and less often, cardiac muscle or neuron channels.	Paralysis.

bind to the Na channel in puffer fish. The source of the 'zombie' state in the voodoo religion has been attributed, controversially, to the use (very carefully) of TTX in 'zombie powder'.

TTX has been an important weapon in the armoury of neuroscientists. It is, for example, useful in separating the two components of axonal current (Figure 4.8). The strong binding of TTX to the Na channel plentiful in, for example, the electroplates of the electric eel, has allowed the extraction and purification of the channel and its subsequent thorough examination. The puffer fish also has the smallest genome known among vertebrates. Unlike the human genome it contains little 'junk DNA'. Thus the study of the less complicated puffer fish genome could help to identify human genes and the DNA sequences that govern their activities.

(2)

(2) *Batrachotoxin (R = CH₃); Homobatrechotoxin (R = C₂H₅)*

Scorpion Suicide by Self Injection is a Myth

Scorpions impale their prey on pincer-like organs, sting them and then inject them with lethal venom. The venom is a complex mixture of polypeptides (toxins) acting in different ways on both Na and K channels, as well as Cl channels, in excitable and nonexcitable cells. This results in convulsions, loss of consciousness and even death (all scorpion toxins are based on 36–66 amino acid polypeptides). Every year several hundred people in Mexico die from scorpion stings. One of the toxins is charybdotoxin (Table 4.5) from *Leiurus quinquestriatus.* It is a polypeptide that

Table 4.4 *Sodium channel kept open*

Toxin and Chemical	Source	Physiological Effect	Clinical Effect
Batrachotoxins (2) alkaloids	Colombian poison dart frog, genus *Phyllobates*, batrachotoxin, (skin) and New Guinea songbirds *Pitohui dichrous* and *Ifrita kowaldi*, homobatrachotoxin (feathers). Only known example of a bird chemical defence.	Block Na channel inactivation. Long lasting membrane depolarization and repetitive neuron firing, acting on nerve, muscle and heart.	Lethal; most potent known venom to smaller animals, but toxic effect in humans is uncertain.
α- and β-scorpion toxins. 64 and 66 amino acid polypeptides	Scorpions.	Slows Na channel inactivation (α) or interferes with activation (β). Repetitive firing in motor units.	Accelerated respiration, convulsions and eventual respiratory failure.

Table 4.5 *Potassium channel blockers*

Toxin and Chemical	Source	Physiological Effect	Clinical Effect
Apamin 18 amino acid polypeptide. Smallest neurotoxic peptide known.	Bee venom, *Apis mellifera*.	Blocks certain classes of K channels, *e.g.* SK in neurons and in smooth and skeletal muscles. Also blocks Ca channels in heart muscles at 0.1 nM.	Convulsions and respiratory failure in mice.
Charybdotoxin 37 amino acid polypeptide.	Scorpion.	Blocks several K(Ca) and voltage gated K channels.	Convulsive paralysis.
Dendrotoxins 59 amino acid polypeptide.	Eastern green mamba snake.	Blocks a variety of K channels.	Convulsant.

Table 4.6 *Calcium channel blockers*

Toxin and Chemical	Source	Physiological Effect	Clinical Effect
Calciseptine 60 amino acid polypeptide.	Black mamba snake.	Selectively blocks L-type Ca channels.	Relaxes smooth muscles and inhibits cardiac muscle contractions. Deadly to humans.
ω-Conotoxins (large family) 24–27 amino acid polypeptides.	Fish-eating cone snails (see also μ-conotoxins).	Various subtypes block different Ca channels in vertebrate neurons. Block neuromuscular transmission.	Paralysis in small mammals and fish. In humans, effects range from bee sting pain to death within 4–5 hours.
Grammotoxin SIA 36 amino acid polypeptide.	Chili Rose tarantula.	Block N-, P- and Q-type Ca channels and eliminates glutamate release from synaptosomes.	Mild effect in humans.
ω-Agatoxins (several) polypeptides.	American Funnel-web spider *Agelenopsis aperta*.	Block presynaptic P-Ca channels.	Paralysis in insects and mice.

contains three S–S bonds for rigidity. The toxin blocks the external pore of a K(Ca) channel, as well as voltage gated K channels. Venom from certain North African and American scorpions contains α-toxins, which bind to Na channels keeping them open and prolonging the action potentials in both nerves and muscles.

Significantly, it has been shown that venom from the scorpion *Androctonus australis* is pharmacologically *ineffective* on both Na and K channels isolated from that species, thus ruling out sensitivity to its own toxins! Venom had no effect on the action potential recorded in the nerve cord axons from *A. australis*, but greatly prolonged the action potential in the crayfish giant axon. The toxins, often polypeptides, are easily proteolysed and are therefore harmless when the predator eats the prey!

- *Toxin Resistance.* The ability to resist its own toxin shown by puffer fish and scorpions has been mentioned above and must be widespread amongst venomous creatures. Another example is the western Colombian frog *Phyllobates terribilis*, which is insensitive to the toxin that it secretes (about 1 mg per frog) from its skin (Table 4.4). This insensitivity is due to a modified binding site on one of its sodium channels. Interestingly, it is still vulnerable to the similarly acting toxins veratridine and grayanotoxin that are found in plants (next section). As an aside, poisonous frogs do not usually synthesize their toxins but rather obtain them by ingesting arthropods that contain them. A modification in the voltage gated sodium channels in muscles, and possibly in neurons, in the western North American garter snake (*Thamnophis sirtalis*) allows it to include the California newt (of the salamander genus *Taricha*) in its diet! The newt has TTX in its skin, which normally acts as a very strong deterrent to predators.
- *Plant Derived Neurotoxins.* Many lipophilic alkaloid toxins act like batrachotoxin but are found in plants. They bind to Na channels to slow or block inactivation and interfere with nerve firing. The plant toxins include the alkaloids veratridine (lily genus *Veratrum*), grayanotoxins (rhododendron leaves) and aconitine (monk's hood, a blue wildflower). Pyrethrins are organic esters derived from chrysanthemum and are natural insecticides that prolong Na channel activation in nerve axons. They cause lethal convulsive paralysis in insects and have been used by humans as stimulants and aphrodisiacs. The structures of all of these compounds are known.
- *Marine Toxins.* Marine toxins are likely to be more of a problem to humans than are toxins from bites or stings. They are responsible

worldwide for very many illnesses and not a few deaths. In 1987, 26 people died as a result of shellfish poisoning from marine toxins accumulated by clams from phytoplankton growing off the Pacific coast of Guatemala. Ciaguatera is poisoning caused from eating certain fish that live near coral reefs, especially in the Caribbean sea and the tropical Pacific ocean. More than $20\,000$ people suffer annually from ciguatera seafood poisoning. The severe digestive, neurological and cardiovascular disorders it causes are however, rarely fatal. Ciguatoxin (3), one of the toxins responsible, acts by opening voltage gated Na channels by binding at the same site as brevetoxins and acting like a bactrachotoxin. Ciguatoxin contains thirteen 5–9 membered rings and thirty stereogenic centres. A close derivative, ciguatoxin CTX3C (3), has recently been synthesized and this will provide a supply (not readily available until now) for the preparation of antibodies to ciguatoxin to use in detecting these toxins.

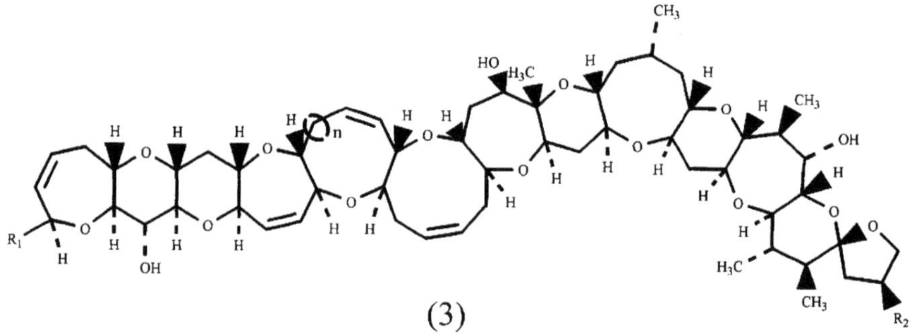

(3)

(3) *Ciguatoxin* $R_1 = -CH=CHCH(OH)CH_2OH$, $R_2 = OH$, $n = 0$; *CTX3C* $R_1 = R_2 = H$, $n = 1$

4.6 LOCAL ANAESTHETICS

We have seen that toxins that interfere with ion channel behaviour can lead to calamitous results. However, the deliberate blocking of a channel (but in moderation!) by synthetic agents can have beneficial effects. Local anaesthetics are generally secondary and tertiary amines [*e.g.* lidocaine (4) and procaine (novocaine) (5)] that exist in the partially protonated form at physiological pHs. This allows them to both penetrate the axonal membrane and block the Na channel, primarily from the inside, near segment 6 of domain IV (Figure 3.2). A local injection at high concentration next to the nerve branch carrying fibres from the region to be anaesthetized reversibly blocks Na membrane channels. This inhibits impulse generation and propagation in nerves, which is very helpful in sensory nerves when painful surgery is required. Since the anaesthetics work on nerves in all excitable tissues, *local* application is necessary and steps must also be taken to impede overall body distribution. Lidocaine and other voltage-dependent Na channel blockers are also used to treat neuropathic pain such as facial neuralgia. Neuralgia probably originates from an increased number of Na channels in sensory nerve fibres and thus leads to spontaneous action potentials in peripheral nerves.

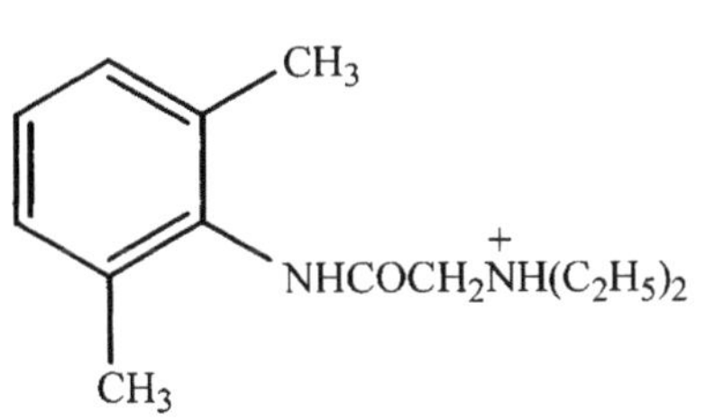

(4) 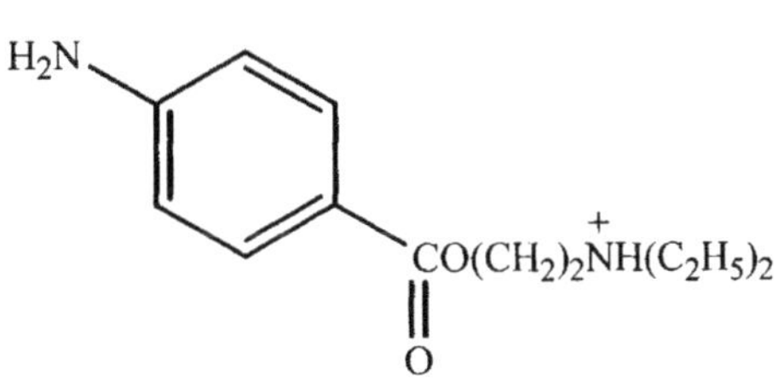   (5)

(4)(5) *Procaine (5), discovered in 1909, is a simplified analogue of cocaine, which is known to have local anesthetic properties. Further modification produced lidocaine (4), which has enhanced stability towards hydrolysis and enzyme catalysed degradation*

Lidocaine and procaine are also used for blocking Na channels that may be responsible for ventricular tachycardia (abnormal increased heartbeat) leading to the irregular pumping of blood.

4.7 SUMMARY

Sodium, potassium and calcium ions play a large part in the transfer of information (as electrical signals) from one part of a nerve (neuron) to another. The movement of these ions through channels (Chapter 3) in the membrane of the neuron is central to the process of setting up the signal (action potential), which is discussed qualitatively. The consequences of neuron malfunctions are considered. Defective ion channel functions can arise from mutations that lead to perturbations in the flow of ions (ion channelopathies). A number of clinical disorders can result. The disturbance can also be external. Neurotoxins in venoms from a variety of species can interfere with the actions of ion channels, again with serious consequences.

Intercellular Signalling

The point at which the axon of the neuron ends is swollen into the *synaptic knob* (Figure 4.3). It is here that the nerve impulse (action potential) temporarily terminates, and it is from here, that information (signal) must be transferred. This can be to another neuron (usually, but not always, at the dendrite) or to a muscle or gland. The gland is a set of epithelial cells that secrete a hormone, mucus, *etc.* for use in and outside the body.

There may be near contact (only a 2 nm or so gap) between the communicating cells using gap junctions (3.9), in which case 'electrical transfer' can be direct. Alternatively, and quite importantly, there may be a small space (synaptic cleft) that must be traversed. In many animals there are more than fifty neurotransmitters that are used in this indirect transfer. Usually sufficient neurotransmitter is released by many synaptic events to depolarize the postsynaptic cell membrane, that is make the membrane potential more positive. If the graded potential generated reaches threshold (4.4) a postsynaptic action potential can be triggered, which can allow the information transfer to proceed.

Therefore we are concerned in this chapter with:

- The process of neurosecretory release from the synapse (calcium ion promoted exocytosis).
- The chemical nature of the neurosecretions.
- Neurosecretory manipulation by the target cell.

5.1 EXOCYTOSIS

At this stage, Ca^{2+} ions take over from Na^+ and K^+ ions as the principal ions participating in neuron function. It was recognized as early as the 19th century that Ca^{2+} ions in the extracellular medium were essential for synaptic operation. Later it was shown that if the Na and K channels in the presynaptic membrane were blocked, synaptic transmitter release was still promoted provided Ca²⁺ ions were present.

In response to the arriving membrane depolarization, Ca^{2+} ions flood into the knob (Figure 5.1) through Ca channels that have opened. Neurosecretors are released from synaptic vesicles in the knob. These can be neurotransmitters, neurohormones, digestive enzymes or even mucus(!). Calcium ions effect this release by promoting fusion of synaptic vesicles with the plasma membrane by a still poorly understood mechanism (2.5.3). Active zone material, an aggregate of proteins, probably provides a scaffold for the docked synaptic vesicles at the presynaptic plasma membrane and the Ca channels anchored in the membrane. There is increasing evidence that exocytosis might also be regulated by the release of Ca^{2+} ions from stores in the ER, which is in close contact with the plasma membrane and associated secretory vesicles. Calcium-independent regulation of secretion in certain cells also appears to operate.

Some idea of the concentrations of Ca^{2+} ions that can effect transmitter release comes in part from experiments using 'caged' calcium (1.5.4). The synaptic terminal of a rat auditory brainstem was loaded with the DM-n calcium complex and a Ca^{2+} ion indicator and was laser photolysed to break down the DM-n Ca^{2+} complex and release Ca^{2+} ions. An increase in $[Ca^{2+}]_i$ of only 1 μM stimulates the release of transmitter (glutamate) as indicated by an excitatory postsynaptic current. A brief spike of 10 μM $[Ca^{2+}]$ is sufficient to reproduce the physiological effect. It has been shown that there is not a uniform distribution of Ca^{2+} ions following the depolarization, but rather microdomains ($<$1 μm²) of high $[Ca^{2+}]$,

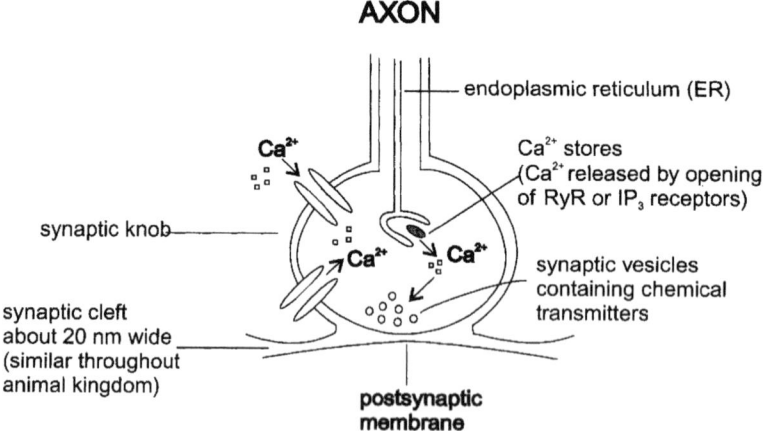

Figure 5.1 *The role of Ca²⁺ ions in exocytosis. Initiated by Ca²⁺ ions (□), neurosecretors are released from synaptic vesicles (○) on to the postsynaptic membrane*

200–300 μM, are generated during exocytosis. As might be anticipated the Ca channels are near the transmitter release sites, closely opposed to the postsynaptic fibre (Figure 5.2).

Which voltage gated Ca channels (3.3.3) are associated with the release of neurotransmitters? As might be anticipated, different channels (and even multiple channels) are used depending on the neuron types. For example only Ca^{2+} ion influx through N- and P-type voltage gated Ca channels (Table 3.1) triggers acetylcholine (ACh) transmitter release in a neuroneuronal synapse in the buccal ganglion of *Aplysia californica*. Another Ca channel also present (L-type) does not participate. This was shown by demonstrating that dihydropyridines (Ca (L) channel blockers) do *not* modify neurotransmitter release. Transmitter release is extremely sensitive to extracellular Ca^{2+} ion concentrations, much more so than to other bivalent alkaline earth ions ($Ca^{2+} > Sr^{2+} > Ba^{2+}$). A pathological rise in serum Mg^{2+} ion concentration can inhibit transmitter release, *e.g.* of ACh, and thus interfere with neuromuscular transmission (2.3.1).

5.1.1 Endocytosis

In order to maintain normal synaptic transmission, the small vesicles fused to the membrane during exocytosis must be retrieved in the process known as endocytosis (2.5.3). The latter process is difficult to examine, but evidence strongly suggests the involvement of Ca^{2+} ions here also. The depolarizing bipolar cell (Figure 7.6B) from goldfish retina has a large synaptic terminal. This allows the examination of exocytosis and endocytosis by the simultaneous determination of Ca^{2+} ion fluxes (using Fura-2) and time resolved capacitance measurements in a patch clamped isolated synaptic terminal. A membrane voltage change (-60 mV to 0 mV) initiates a jump in capacitance arising from an increase in membrane surface area as vesicles fuse (exocytosis). The restoration of the capaci-

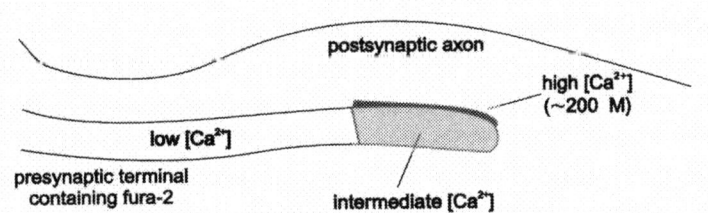

Figure 5.2 *Visualization of Ca^{2+} ion levels during transmitter release in the squid giant synapse. At rest only low [Ca^{2+}] is observed throughout. The indicators fura-2 or aequorin were used and a fluorescence microscope was linked to a computer for digitization of [Ca^{2+}] images*

tance to the resting level corresponds to membrane retrieval and the recycling of vesicles (endocytosis). The processes are complex, but the changes in Ca^{2+} ion concentration mirror those of the capacitance.

5.1.2 Neurotoxins That Target Exocytosis

The disturbance of the neuroexocytosis process is the basis of action of certain chemicals called neurotoxins, both bacterial and non-bacterial in origin. Some of the most lethal toxins known interfere with exocytosis (Table 5.1). They can enhance, but (more usually) inhibit neurotransmitter release and thereby block transmission of the nerve impulse.

Neurotoxins like the botulinum and tetanus and LTX, which target exocytosis, are useful in research for the identification of the synaptic proteins that are implicated in exocytosis. Later we will meet neurotoxins that act postsynaptically.

5.1.3 Clostridial Toxins

Bacteria of the genus *Clostridium* are widely distributed in soil. Some species, including *Cl. botulinum*, *Cl. tetani* and *Cl. perfringens*, produce lethal toxins. The *Cl. botulinum* organism was originally isolated from contaminated sausage (*L. botulus*, sausage). Botulinum toxins are the most potent known poisons, three nanograms being lethal to a human being. These toxins are ~150 kDa polypeptides that form a heavy chain (~100 kDa) that binds to the extracellular membrane receptor (in the presence of Ca^{2+} ions) and aids the entry into the nerve terminal of the light chain (~50 kDa), which is a Zn endopeptidase and is the toxic component. After the light chain has broken off, the component zinc

Table 5.1 *Neurotoxins that target exocytosis.*

Toxin and Chemical Nature	Source	Physiological Effect	Clinical Effect
Botulinum toxins (several) polypeptides	*Clostridium botulinum* Responsible for botulism (food poisoning).	Inhibit depolarization- induced transmitter release at the neurotransmitter junction.	Flaccid muscle paralysis in humans and other animals.
α-Latrotoxin (LTX) polypeptide	Black widow spider *Lactrodectus mactans.*	Promotes massive release of transmitter at vertebrate presynaptic terminals.	Paralysis of respiratory muscles.

enzyme destroys proteins critical to exocytosis. Release of ACh from motor nerve endings is thus prevented and muscle paralysis results. Tetanus toxin has a similar structure and action. It acts on spinal inhibitory interneurons, inhibiting the release of glycine, and causing spastic paralysis, *e.g.* of the jaw (lockjaw). The symptoms of tetanus (Gk., *tetanos*; stretched, spasm) poisoning were first described by Hippocrates. In minute doses, botulinum toxin A has been used to treat a variety of diseases involving spasms of the face, neck or eye muscles. These spasms arise from the release of excess ACh, which can be moderated by intramuscular injection of the toxin ('Botox boom').

A particularly nasty toxin (α-toxin) is released from *Cl. perfringens*. It produces gas gangrene (rotting flesh), a condition that was very common in World War I. One of the two enzymes that comprise the toxin is a Zn metalloenzyme with phospholiphase C activity. It also has a phospholipid binding domain, C_2 (1.3.3), implicating binding of Ca^{2+} ions, which probably activate the cell membrane and help to burst the cell wall. This enzyme resembles mammalian counterparts that pass messages into cells and trigger inflammation.

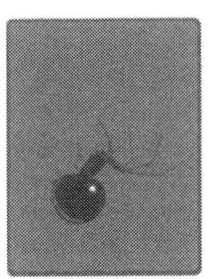

The Black Widow Spider Knows What Ca^{2+} Ions Do!

Many latrotoxins (LTX) found in the venom of the black widow spider target insects. Alpha latrotoxin stimulates (even in picomolar concentrations) massive neurotransmitter (ACh) release from a variety of vertebrate neuronal cells, including those at the neuromuscular junction. This can (but rarely does) lead to death by paralysis of respiratory muscles. LTX may act as a Ca^{2+} ionophore (by facilitating transmembrane transport) or could just cause uncontrolled exocytosis by binding to neurexin in the plasma membrane (2.5.3). The structure of LTX has been determined recently by single particle cryoelectron microscopy. In cryoelectron microscopy a very thin layer of a solution (~1 mg ml^{-1}) of the sample is placed on a carbon coated electron microscopy grid and rapidly cooled in liquid ethane. The resulting solid glass-like sample is placed in a vacuum chamber in the microscope at liquid nitrogen temperature, and examined. Although the resolution is relatively low (~20 Å), overall characteristics can be seen and transient conformational states can be frozen out with a time resolution of only milliseconds. The active tetramer (520 kDa) is formed only in the presence of Ca^{2+} or Mg^{2+} ions. It adheres to the membrane and forms channels there, either for Ca^{2+} ions to enter or perhaps for neurotransmitter to exit (since it can stimulate exocytosis even in the absence of Ca^{2+} ions, although not without Mg^{2+} ions). Either event would stimulate massive exocytosis.

The LTX tetramer binds to a receptor protein, latrophilin, which is a G-protein coupled receptor in the presynaptic membrane. Latrophilin could be activating phospholiphase C signalling and causing IP_3-induced mobilization of intracellular calcium ions, again stimulating exocytosis.

5.2 THE CHEMICAL NATURE OF NEUROSECRETIONS

Neurosecretions are released from the synaptic vesicles, they transverse the synaptic cleft and can either be released into the bloodstream (neurohormones) or bind to specific receptors in the postsynaptic membrane of another neuron, muscle or gland cell (neurotransmitter). The delay between the presynaptic current due to Ca^{2+} ion entry into the synaptic knob and the corresponding postsynaptic response is less than $200\ \mu s$, which strongly suggests that a complex multistep process is not involved in the neurosecretory transfer. Neurohormones are hormones from the nervous system. Many neurohormones are peptides released from the terminals of neurosecretory cells into capillaries. After entering the bloodstream, some neurohormones act directly on somatic target tissues in remote organs. An example is oxytocin (Gk. swift childbirth), a nine amino acid polypeptide (1), which regulates smooth muscle contraction by an unknown mechanism. It is best known for stimulating uterine smooth muscle contraction during childbirth and for releasing milk from mammary glands. Oxytocin was the first biologically active polypeptide to be chemically synthesized. Neurohormones are usually longer lived than neurotransmitters because of the distances over which they operate although they could also function as neurotransmitters.

$$H_2N - G - L - P - \overset{\displaystyle \lceil\!-\!S-S-\!\rceil}{C} - N - Q - I - Y - C$$

(1)

5.3 NEUROTRANSMITTERS

Some neurotransmitters are small gaseous molecules, *e.g.* NO and CO. These are atypical since they do not use the presynaptic/postsynaptic path, but simply diffuse from one cell site to another (Ca^{2+} ions are also involved here, see 6.5.3). Some important neurotransmitters and the receptors they influence are shown in Table 3.3. The significance of the neurotransmitters described in that table will become clear shortly.

When the synaptic transmission must be fast, small neurotransmitters are employed. These include ACh, which is one of the most important neurotransmitters and is the only one used at the neuromuscular junction. Others are amino acids, biogenic amines (decarboxylated amino acids, *e.g.* histamine, norepinephrine) and adenosine derivatives. For slower transmission, for instance a 100 ms response time lasting for anything up to hours, larger neuropeptides and proteins (nearly 100 have been identified so far) are used. These usually act through a second messenger system and the ion channel is opened indirectly. Both fast and slow transmissions use Ca^{2+} ion-induced exocytosis and both could result from the activity of a single neuron.

5.3.1 Fate of Neurotransmitter After Use

When a neurotransmitter has bound to its receptor and carried out its duties (see next section), it can then be taken back, by diffusion or active transport, into the presynaptic terminal, thus ending the transmission. Acetylcholine (as choline after enzyme catalysed hydrolysis), GABA, glycine and amines use a transporter ($2Na^{+}$, $1Cl^{-}$ and 1 neurotransmitter, all into the cell) in the presynaptic terminal. Glutamate uses a $3Na^{+}$ and $1H^{+}$ (in) and a $1K^{+}$ (out) for each glu (in) transporter.

Interference in the Reuptake of Neurotransmitter can be Both Good and Bad!

Fluoxetine (Prozac, 2) inhibits the reuptake of serotonin into the presynaptic membrane. It thereby forces serotonin to remain in the space surrounding the nerve ending. Serotonin plays a role in various brain functions, *e.g.* the mediation of sleep, aggression, *etc.* Its role is suggested by the behaviour of knockout mice that lack the gene for one 5HT receptor and that are uncommonly aggressive. Fluoxetine is a much-prescribed antidepressant that is without serious side effects. However, two other serotonin reuptake inhibitors, sertraline (Zoloft, 3) and paroxetine (Paxil, 4) that are used as antidepressants have been implicated in violent patient behaviour (including murder and suicide).

Any interference in the acetylcholinesterase (AChE) catalysed hydrolysis of ACh to choline and acetate will result in the build up of ACh and the consequent disruption of the nervous and neuromuscular systems, leading to death. Mutant zebrafish that lack acetylcholinesterase activity have impaired muscle fibre development and premature sensory neuron death. Sarin (5) nerve gas (released in the Tokyo underground in 1995), certain insecticides and toxins from cyanobacteria (blue-green algae) act by blocking the activity of AChE. Thus, respiratory muscles are continuously contracted and eventually give up!

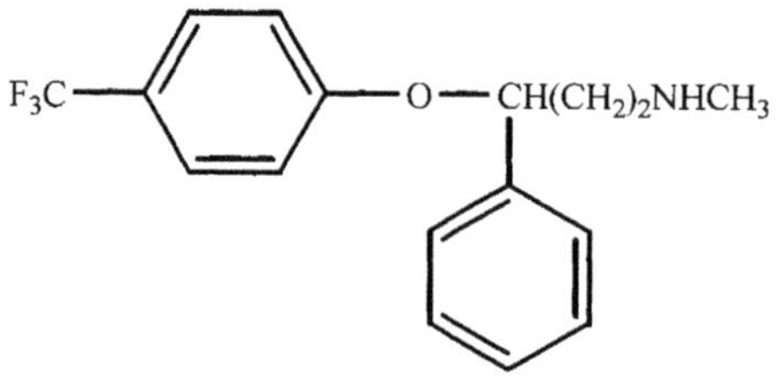

(2)

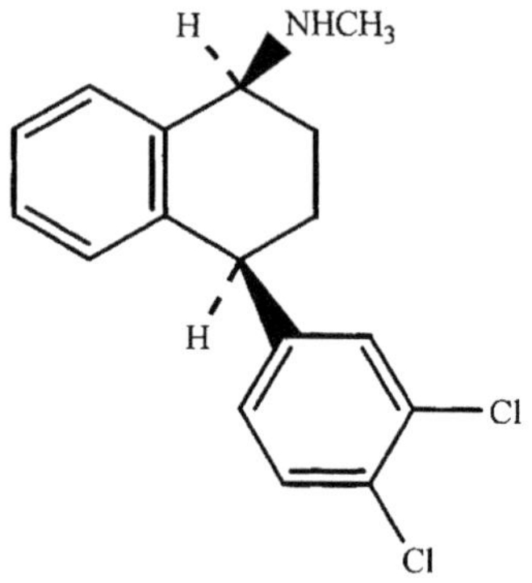

(3)

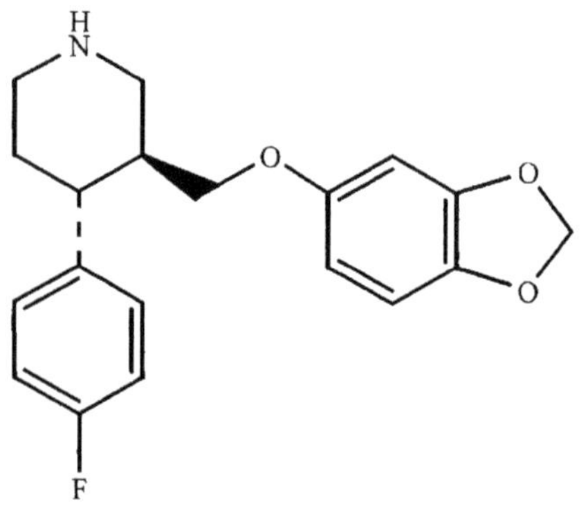

(4)

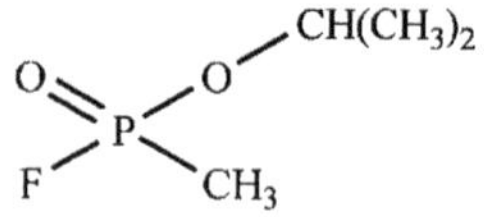

(5)

5.3.2 Manipulation of Neurotransmitter by the Target Cell

The postsynaptic receptor and the type of ion channel it incorporates or promotes determines whether the postsynaptic potential (PSP) is inhibitory

(IPSP) or excitatory (EPSP) – that is, whether the postsynaptic cell is, respectively, less likely or more likely to fire an action potential and continue to transmit the message (Figure 5.3). The neurotransmitter is likely to play a role in determining the conditions that dictate either an IPSP or EPSP response, although a neurotransmitter can be excitatory at some synapses and inhibitory at others. A single neuron can receive thousands of synaptic inputs. It sums the IPSP and EPSP signals and then decides whether or not to fire an action potential. In this way, it plays an important role in the CNS computation processes. Brain circuits for example, need balanced positive and negative signals to function properly.

5.4 INHIBITORY POSTSYNAPTIC POTENTIAL (IPSP)

Consider, for example, a potassium or a chloride channel incorporated in a receptor and ignore the Na channel. Binding of the neurotransmitter is assumed to open the channel and allow the efflux of K^+ ions or the influx of Cl^- ions (Figure 5.4A). In either case, this will lead to hyperpolarization of the postsynaptic membrane potential (see Figure 5.4B) and generation of an action potential will be inhibited. Glycine and γ-aminobutyric acid (GABA) are such inhibitory neurotransmitters. Their corresponding receptors (glycine and $GABA_A$) are very prevalent in CNS synapses. Both are also chloride channels. Binding to ionotropic $GABA_A$ receptors promotes

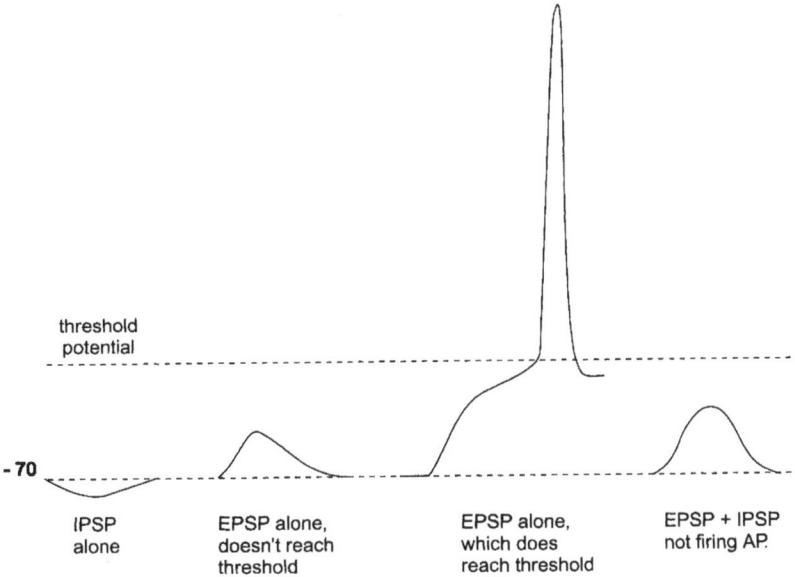

Figure 5.3 *IPSP and EPSP. Inhibitory, excitatory and combination impulses*

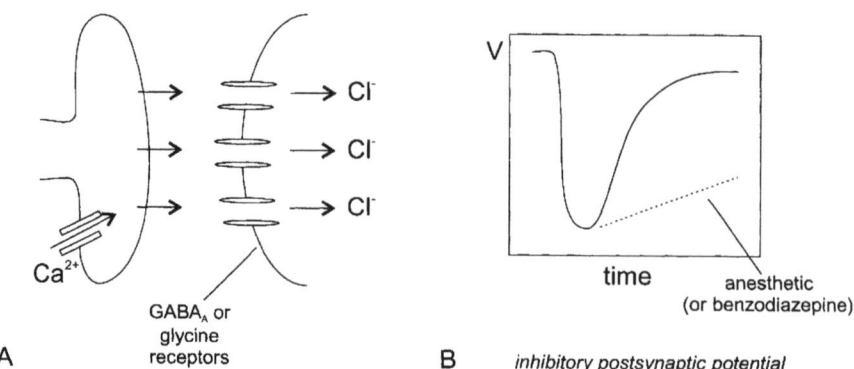

Figure 5.4 *General anaesthetics and sedatives use inhibitory neurotransmitter channels. (A) Receptors bind neurotransmitters and open to allow chloride ions to move across the postsynaptic membrane. (B) The volatile anaesthetics prolong the channel opening and therefore increase postsynaptic inhibition*

a rapid response. Any ligand whose binding either enhances the frequency of, or prolongs, chloride channel opening will increase postsynaptic inhibition (Figure 5.4B). Relaxation, immobility or even unconsciousness results. This is the basis of action of most (but not all) general anaesthetics and sedatives, including alcohol!

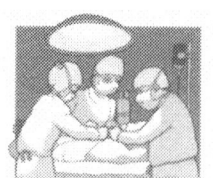

General Anaesthetics and Sedatives Act on Chloride Channels

Both volatile liquid and gaseous general anaesthetics were introduced in the 1840s, but the basis of their action has been puzzling ever since. It is now believed that they function, in part, by binding to Cl^- permeable $GABA_A$ or glycine receptors, and aid ion channel opening (Figure 5.4). The Cl^- ion diffusion across the postsynaptic membrane causes hyperpolarization of the membrane, short circuits response to depolarization input and dampens neuron electrical activity.

Anti-anxiety drugs of the benzodiazepine family (*e.g.* diazepam, (Valium, 6)), introduced in the 1960s, act in a similar fashion by binding to the $GABA_A$ receptor as an agonist. The binding sites for the drugs are different (also for general anaesthetics) than those for GABA (Figure 5.5). As of now there are no known human diseases associated with mutations in genes encoding $GABA_A$ receptor subunits.

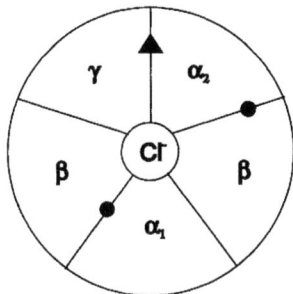

Figure 5.5 *Schematic of a GABA_A receptor. There are many types of GABA_A receptors. A schematic of a common GABA_A receptor is shown. The five subunits, each consisting of four transmembrane domains, form a ring enclosing the chloride channel. The GABA (●) and benzodiazepine drug (▲) binding sites are shown (see Figure).*

(6)

As well as inducing relaxation, benzodiazepines reduce concentration and can cause physical clumsiness and tricks of memory. This occurs because GABA$_A$ receptors abound in the brain. Attempts are being made to develop drugs specific for GABA$_A$ receptor subtypes, which are responsible for reducing anxiety, but without the unpleasant side effects.

As might be expected, any antagonist that blocks GABA$_A$ or glycine activated Cl$^-$ channels will prevent the inhibition of nerve impulses, allowing free rein to neuronal firing. Such an antagonist would act as a powerful CNS stimulant. The alkaloid picrotoxinin (7), contained in plants of the moonseed family, is a non-competitive antagonist of both GABA$_A$ and glycine activated Cl$^-$ channels. Strychnine (8) from the seeds of *Strychnos* species is a competitive antagonist of glycine gated Cl$^-$ channels. Both are powerful convulsants.

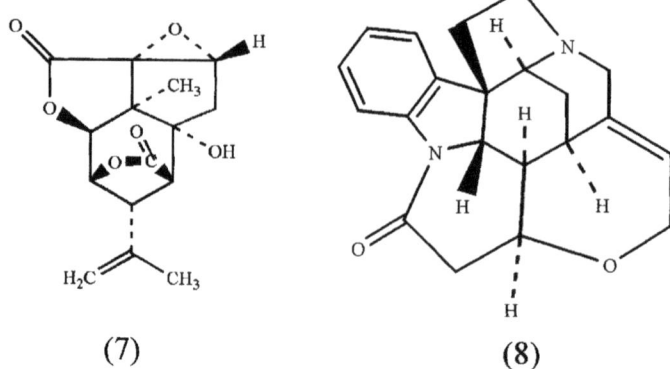

(7)   (8)

Van Gogh

Absinthe Makes the Neurons Wilder!

The toxin α-thujone (9) blocks brain GABA$_A$ receptors. The toxin is found in absinthe, a green potent aniseed-flavoured liqueur. The liqueur is made from a shrubby plant (wormwood) and turns milky white when added to water. It was the national drink of France in the late 19th century, but its reputation for (apparently) inducing hallucinations or criminal or insane acts (van Gogh's suicide has been attributed to absinthe drinking) led to its prohibition in most countries early in the 20th century. Thujone is also found in plants of the conifer family that includes *Juniperus* (a source of gin flavour) and a distant relative, hemlock.

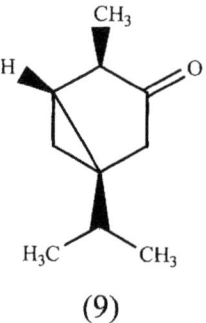

(9)

The essential roles of the GABA$_A$ or of the glycine receptors in the transmission of inhibition are demonstrated by the behaviour of mutant forms. A frog-egg cell becomes unresponsive to an anaesthetic when a single amino acid change is made in the 2000 or so that make up its

GABA$_A$ proteins. Mutations in the glycine receptor can simulate antagonists by reducing glycine-activated currents. These give rise to the very rare, so called, startle diseases. Since glycinergic interneurons have a major role in normal spinal cord reflexes and muscle movement, the malfunction leads to facial and limb spasms and even falling down after being startled (*e.g.* a sudden noise). The condition has been observed in humans and mice. In poll Hereford cattle and Peruvian Paso horses, reduced glycine receptor expression in the spinal cord leads to myoclonus, resulting in calves born with dislocated hips.

5.5 EXCITATORY POSTSYNAPTIC POTENTIAL (EPSP)

The direction of Cl^- ion movement through the GABA- or glycine-promoted opened channel will, of course, depend on the relative concentrations of chloride ions inside and outside the cell. A lower intracellular $[Cl^-]$ will set up the conditions for influx of Cl^- ions and hyperpolarization on neurotransmitter stimulation as described above for IPSP. In early life, the chloride ion concentration in rat hippocampal cells, for example, is high and consequently there is a Cl^- ion efflux on stimulation. This leads to depolarization of target cells and an excitatory postsynaptic potential (EPSP). About a week after birth the expression of a K^+/Cl^- cotransporter results in Cl^- ion loss from the cell. Now there is a Cl^- ion influx on stimulation and hyperpolarization and therefore a change to IPSP from EPSP occurs. It is surmised that the early depolarization conditions are required for Ca^{2+} ion entry into the cell, which is used for gene expression.

An important example of an excitatory transmitter is ACh binding to *n*AChR at the neuromuscular junction, which is a much-studied prototype. In vertebrate species, the synapse between motor neurons and skeletal muscle is always excitatory. There are many branches at the end of the motor neuron axon. Each branch forms a single junction with a muscle fibre by lying embedded in grooves on the fibre surface. Every nerve impulse that reaches the motor neuron ending releases about 300 vesicles, which contain approximately 10^4 molecules of ACh per vesicle. The neurotransmitter binds to the *n*ACh receptor (3.5.1) (about 10^4 per μm^2) in the motor end plate, which is in the plasma membrane of the muscle cell near the synaptic cleft.

The *n*ACh receptors are ligand gated cation channels that allow Na^+ ions to enter and depolarize the postsynaptic cell membrane. This local depolarization of the end plate (end-plate potential, EPP) activates nearby voltage gated Na channels and spreads action potentials across the surface of the muscle fibre and into the muscle fibre system of membranes. This is an important step in muscle contraction, which will be discussed in full later. Each time an action potential reaches the nerve terminal of an α-motor neuron, enough transmitter is released to initiate an action potential on the skeletal muscle membrane. α-Motor neurons are large spinal neurons that innervate extrafusal skeletal muscle fibres in vertebrates. However, at other excitatory synapses (*e.g.* glutamate binding to its receptor at the squid giant synapse) the neuron must fire repeatedly so that enough transmitter is released to generate an action potential at the postsynaptic membrane. This is because much less neurotransmitter is released here than in an α-motor neuron. As a result fewer ion channels are opened and a single EPSP may be only ≈ 0.5 mV, much less than the 15 mV or more required to reach the threshold potential. Most excitatory synapses terminate near dendrites, whereas inhibitory synapses terminate closer to the axon hillock since they need to be more effective (see Figure 4.2).

Finally, the efficiency of synaptic transmission is higher the more frequently it is used. This property is termed *plasticity* and is believed to be the reason the synapses are so important in learning and memory.

5.6 NEUROMUSCULAR DISEASES

Neuromuscular transmission, which has just been discussed, is the target of neuromuscular diseases. Mutations in critical components of the neuromuscular junction, or autoimmune responses to them, can cause abnormal muscle weakness and fatigability (myasthenia). Congenital myasthenic syndromes have been found in which there are defects in almost all the steps involved in neuromuscular transmission. Each produces a characteristic muscular abnormality. For example, a reduced number of voltage gated calcium channels in the motor nerve terminal as a result of the production of antibodies to these channels by the body leads to the production of fewer synaptic vesicles and release of less ACh (Lambert–Eaton myasthenic syndrome). A number of different mutations in genes encoding for the subunits of n-AchR can have several effects. They can reduce the number or affinity of n-AchRs, the rate at which they open and close and cause spontaneous opening even in the absence of ACh! These mutations result in muscle weakness and fatigue (congenital myasthenic weakness).

5.7 NEUROTOXINS THAT TARGET POSTSYNAPTIC RECEPTORS

Postsynaptic receptors are the targets of a number of toxins. Deadly neurotoxins attack the n-acetylcholine receptor (n-AchR), and interfere

Table 5.2 *Neurotoxins that target n-acetylcholine receptors*

Toxin and Chemical Nature	Source	Physiological Effect	Clinical Effect
Histrionicotoxin alkaloid (10)	Columbian frog (arrow poison)	Noncompetitive channel blocker.	Flaccid paralysis. Death by asphyxiation.
D-tubocurarine alkaloid (11)	Cinchona plant bark. Active ingredient in curare, used in native South American blow darts.	Competitive channel blocker, by binding to two ACh binding sites. The N^+–N^+ distance (1.4 nm) in the toxin is critical to its action.	Flaccid paralysis.
α-Bungarotoxin polypeptide	Banded krait snake.	Competitive channel blocker.	Flaccid paralysis.

with the proper functioning of the neuromuscular junction by blocking the associated channel (Table 5.2). In the most severe cases this can lead to respiratory muscle arrest. Modification of the *n*-AchR can alter its sensitivity to toxins. Other receptors are also the targets of neurotoxins (Tables 5.3–5.5).

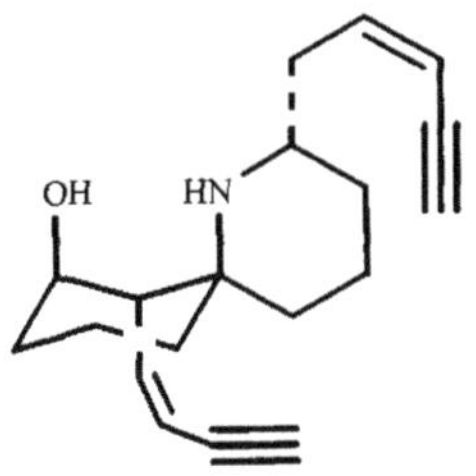

(10) 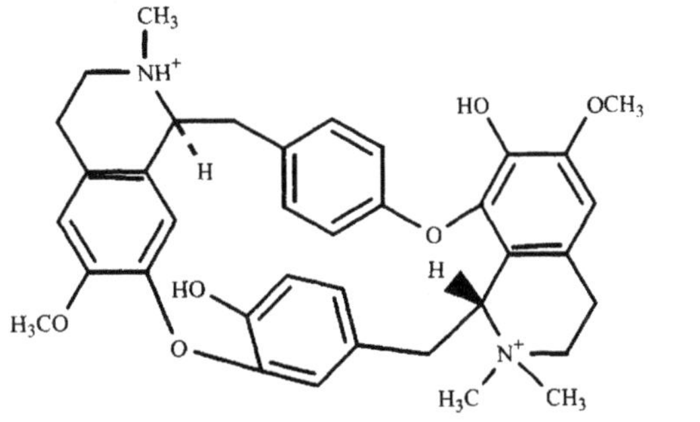   (11)

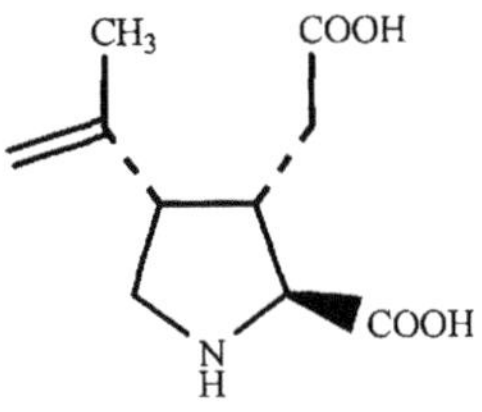

(12) 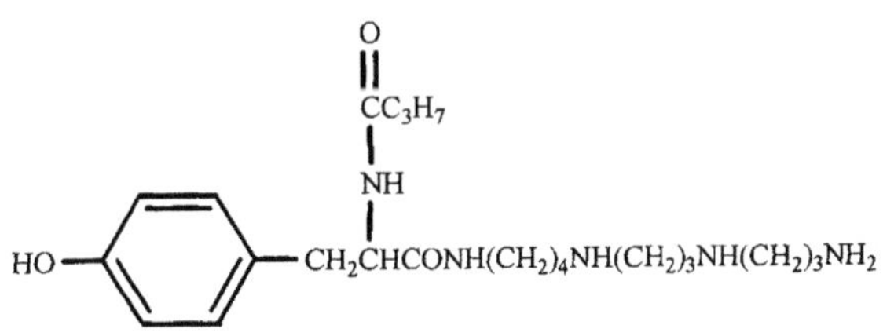   (13)

Table 5.3 *Neurotoxins that target glutamate receptors*

Toxin and Chemical Nature	Source	Physiological Effect	Clinical Effect
Kainic acid (12) cyclic amino acid	Red algae.	Agonist for AMPA subtypes of glutamate receptor. Leads to release of glutamate and aspartate from nerve terminals.	Induces seizures.
Conantokins (Philipino, antokin; sleepy) Conantokin G is called sleeper peptide and has the sequence: G-E-gla-gla-L-Q-gla-N-Q-gla-L-I-R-gla-K-S-N, where gla is γ-carboxyglutamate (2.5.5).	Cone snails.	Noncompetitive antagonists of NMDA receptors (glu and gly are coagonists of this receptor).	Induces sleep in young mice and hyperactivity in older mice.
Philanthotoxin acylpolyamine (13)	Solitary wasp venom, *Philanthus triangulum.*	Noncompetitive channel blocker in insect muscles (blocks voltage gated Ca^{2+} channels at high concentration).	Flaccid paralysis in insects.

Honeybees are a Favourite Snack of the Solitary Wasp

The venom of the solitary digger wasp contains a number of low molecular weight toxins. One of the most investigated, philanthotoxin 4.3.3 (13), acts by blocking neuromuscular transmission in injected insects. It inhibits cationic channels in glutamate receptors in the postsynaptic membrane of insect skeletal muscles. It paralyses worker honeybees (a favourite prey) that then become the host for the wasp larvae. In contrast, the bees (when not being paralyzed!) and social wasps kill their prey directly. A honey bee sting contains, in addition to apamin (Table 4.5), the more abundant and pain causing mellitin. This is a very surface active, 26 residue polypeptide, with marked cytolytic activity (*i.e.* destruction of the outer membrane and breakdown of a wide variety of cells).

Table 5.4 *Neurotoxins that target GABA receptors*

Toxin and Chemical Nature	Source	Physiological Effect	Clinical Effect
Muscimol alkaloid (14)	Mushroom, *Amanita muscaria.*	Competitive agonist of GABA$_A$ activated Cl$^-$ channel.	Hallucinogenic.
Bicuculline alkaloid (16)	Dutchman's breeches plant.	Competitive antagonist of GABA$_A$ activated Cl$^-$ channel. Reduces frequency and duration of opening.	CNS convulsant.

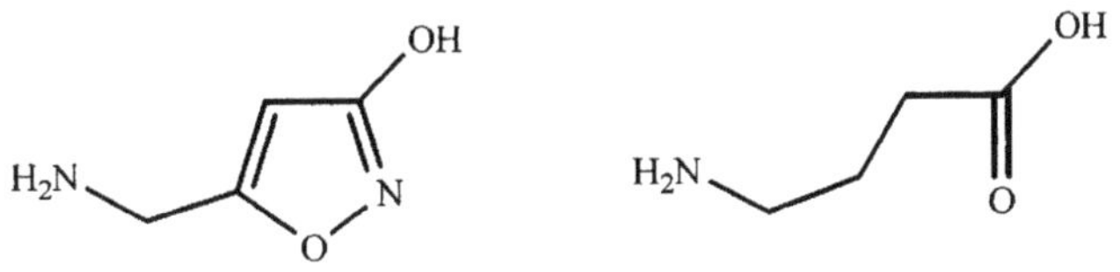

(14)   (15)

(14)(15) Muscimol (14) is bioisosterically derived from GABA (15)

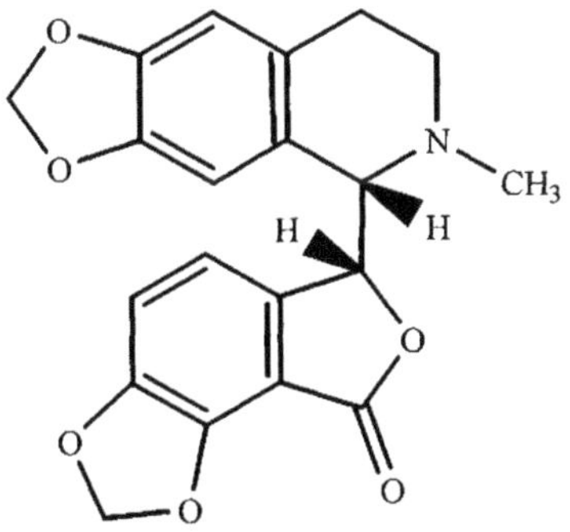

(16)

(16) Occurs naturally as the D-form

More than 100 snake toxins have been isolated from their respective venoms. These neurotoxins can also be used, in carefully controlled amounts, as muscle relaxants. D-tubocurarine is clinically useful as a paralytic agent. Controlled intravenous injection of the compound relaxes

Table 5.5 *Neurotoxins that target glycine receptors*

Toxin and Chemical Nature	Source	Physiological Effect	Clinical Effect
Strychnine (8) Heterocyclic alkaloid	*Strychnos nux-vomica* seed.	Competitive antagonist of glycine gated Cl⁻ channel.	Motor convulsions and progressive paralysis.

muscle fibres and makes them easier to handle during surgery. Similarly, succinylcholine (17) (suxamethonium) is a drug that acts as an ACh agonist that persists and produces prolonged end plate depolarization and drawn out muscle relaxation for easier manipulation during surgery. Interestingly it has the (apparently) crucial N^+-N^+ distance of about 1.4 nm observed in D-tubocurarine. Some snake toxins have also been useful in differentiating among various receptor (channel) types and in the definition of their roles in neuronal processing and synaptic transmissions.

$$H_2C \cdot \overset{\overset{\displaystyle O}{\displaystyle \|}}{C} \cdot O(CH_2)_2 N^+(CH_3)_3$$
$$|$$
$$H_2C \cdot \underset{\underset{\displaystyle O}{\displaystyle \|}}{C} \cdot O(CH_2)_2 N^+(CH_3)_3$$

(17)

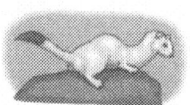

How Can the Mongoose Fight the Snake?

The region of the *n*AChR that contains the binding site domain for neurotoxins differs by only three amino acids in both the primitive snake (sand boa) and the mongoose compared to that from all other animals, *e.g.* the mouse, which is vulnerable to the toxin. Snake α-neurotoxins, including α-bungarotoxin, can no longer bind to the toxin receptor and therefore cannot compete with ACh. This confers neurotoxin resistance on the snake and thus it can't poison itself (compare scorpion, 4.5.2), nor can it injure the mongoose. Now it is a fair fight!

5.8 SUMMARY

Signalling between cells is necessary. This can be direct between cells in near contact (gap junctions) or across a gap (synapse). The latter process was emphasized, that is signalling between a neuron and another neuron, muscle or gland. This is effected by the release (called exocytosis) of neurosecretions (neurohormones and neurotransmitters) from the presynaptic terminal to the receptors on the target cell. Exocytosis is invariably promoted by Ca^{2+} ions. The manipulation of the neurotransmitter by the target introduced the concepts of inhibitory postsynaptic potential (IPSP) and excitatory postsynaptic potential (EPSP). The postsynaptic cell is less likely (IPSP) or more likely (EPSP) to fire an action potential and allow the message to continue. The nerve–muscle synapse and neuromuscular diseases were emphasized. Neurotoxins from plants, bacteria and animals that can target the exocytosis process or postsynaptic receptors occur extensively.

CHAPTER 6

Muscle

6.1 MUSCLE TYPES

There are three major types of muscles (Table 6.1). All can be stretched
and contracted under stimulus.

All muscle contractions take place by the sliding of thick and thin
filaments past each other. They rely on the proteins actin and myosin as
the star performers, MgATP as an energy source and there is a leading role
for Ca^{2+} ions and supporting roles for Na^+ and K^+ ions. The detailed
mechanisms differ, especially between smooth muscle and the two other
types.

6.2 SKELETAL MUSCLE

The vertebrate striated muscle is composed of transverse striped fibres
resulting from the fusion of many muscle cells (Figure 6.1). An electrically
excitable membrane (*sarcolemma*) encloses these. A cytoskeleton support
system prevents tearing of the membrane during contraction and
relaxation.

Absence or Reduction of Dystrophin in Muscle Leads to Dystrophy

Dystrophin is a large protein reinforcing muscle
fibres and preventing them from tearing during
excitation and contraction, although its precise
function is still poorly understood. Duchenne muscular dystrophy (DMD) is due to a defect in the
gene encoding for dystrophin and causes tears in
the plasma membranes surrounding skeletal and
cardiac muscle cells. This allows the entry of Ca^{2+}
ions from the interstitial fluid, which interferes with
Ca^{2+} ion–protein digestive enzyme processes and causes eventual

Table 6.1 *Characteristics of three forms of muscles*

Skeletal	Cardiac	Smooth
Large number of parallel fibres giving a striated or striped appearance.	Large number of branching cells forming a complex network, still striated. Possesses characteristics of skeletal and smooth.	Small cells running in different directions, nonstriated. Part of organ tissue, *e.g.* bladder.
Supports and moves attached skeleton under voluntary (conscious) control by nervous system.	Propels blood through circulatory system in vertebrate hearts. Involuntary control (*e.g.* while sleeping).	In hollow internal organs, *e.g.* moves solids and liquids through gastrointestinal system from mouth to rectum. Involuntary control.
Contraction initiated by nerve impulse using innervation of each fibre.	Initial contraction spreads to entire organ *via* gap junctions (intrinsic rhythmically).	Contraction initiated by nerve impulses, hormones and neurotransmitters.
Small T-tubules, well developed SR systems.	Large T-tubules and reasonable SR systems.	Poor T-tubules and limited SR systems.
Contraction activated by Ca^{2+} ions from SR; binding to troponin key step.	Contraction activated by Ca^{2+} ions from SR and extracellular; binding to troponin key step.	Contraction activated by Ca^{2+} ions from SR and (vital) extracellular; activating myosin (*via* calmodulin) key step. No troponin present.
Sliding thick and thin filaments with slow—rapid cycling of cross bridges. ATP energy source.	Sliding thick and thin filaments with slow cycling of cross bridges. ATP energy source.	Sliding thick and thin filaments with slow cycling of cross bridges, allowing prolonged contraction. ATP energy source.

local necrosis of the muscle. It is a male defect only and affects about 1 in 4000 newborn babies in the US annually. Individuals with DMD usually die before the age of 30 from respiratory or cardiac muscle failure.

There is an interlacing network wrapped around the muscle fibre comprised of the sarcoplasmic reticulum (SR) and a transverse tubular system. Muscle fibres are held together by a connective tissue made up of different proteins, but mainly of collagen. This connective tissue aids in the transmission of fibre movement to tendons, which constitute a solid connection joining the muscle to bone.

6.2.1 Final Events Leading to Skeletal Muscle Contraction

Skeletal muscles are important effector organs (see Figures. 4.1 and 4.2). The neuromuscular junction and the propagation of an action potential to the sarcolemma have already been discussed. The final events leading to muscle contraction will now be outlined. Calcium ions in particular are

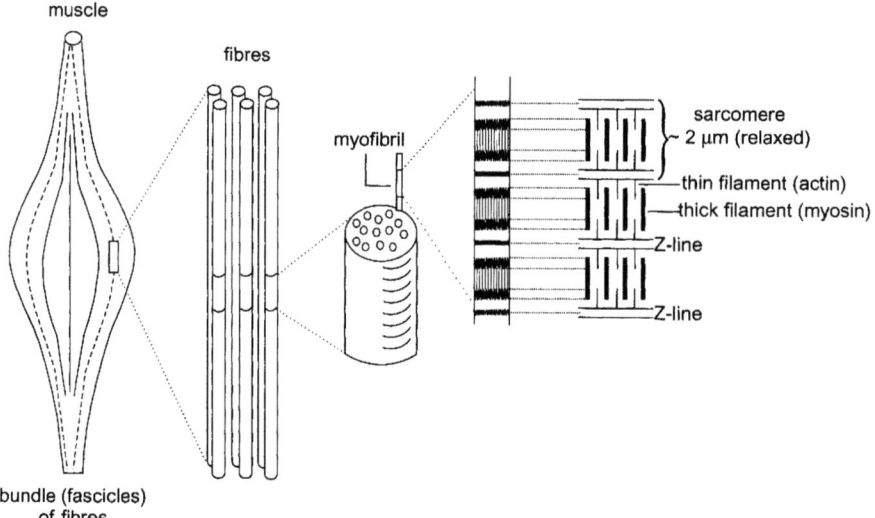

Figure 6.1 *Vertebrate striated muscle. Skeletal muscle consists of bundles of muscle fibres. Each muscle fibre contains the nuclei, mitochondria and SR surrounding numerous myofibrils. Myofibrils are highly ordered assemblies of thick and thin filaments that give rise to the characteristic lighter and darker bands*
Reprinted with permission from Cambridge University Press from K. Schmidt-Nielsen, *Animal Physiology*, 1997.

required everywhere! The objective is to produce an elevated Ca^{2+} ion concentration (10–100-fold over the resting value) near the myosin and actin filaments in the muscle. The sequence of events that link the action potential in the muscle cell plasma membrane to the contractile sliding machinery in the muscle is called *excitation–contraction* (EC) *coupling*.

The action potential, which has reached the sarcolemma, is now transmitted to the interior of the muscle cell with the help of excitable membranes called the T-tubules. These are special invaginations of the plasma membrane situated at regular intervals along the muscle fibre and are open to the extracellular space (Figure 6.2). Electrical signals proceeding down the T-tubule use two different kinds of calcium channels to promote the release of Ca^{2+} ions from the sarcoplasmic reticulum (the SR is equivalent to the ER in other cells):

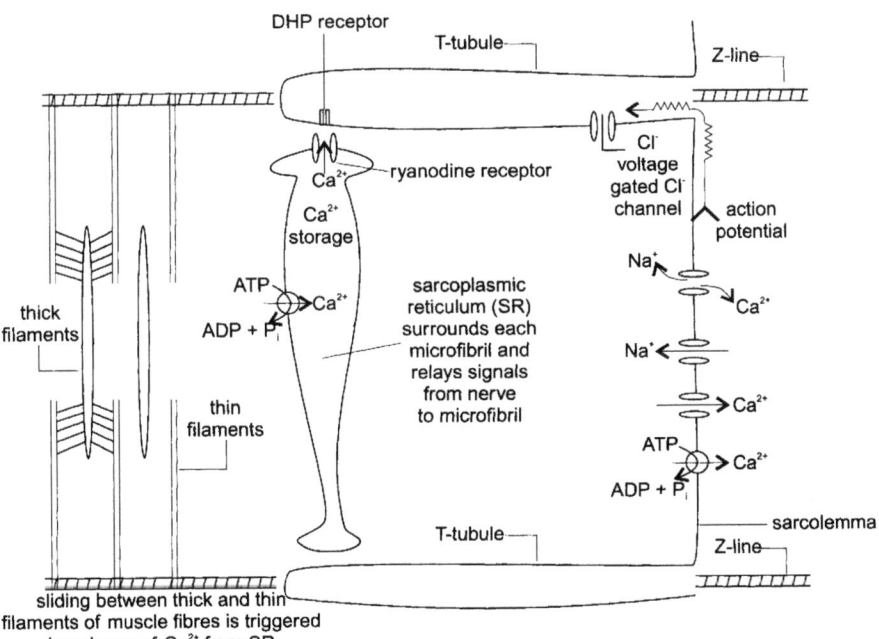

Figure 6.2 *The conversion of action potential into contraction of muscle is depicted. Not all of the ion flows shown are equally important. In addition, there are the ubiquitous Na^+/K^+ATPase pumps on T-tubules and sarcolemma. T-tubules flattened between two SR terminal cisterns form a triad. Using voltage clamped frog single fibre and fluo-3 dye with confocal imaging, it was shown that Ca^{2+} ion 'sparks' originate at the triad. Local concentration increases of $[Ca^{2+}]_i$ in quiescent rat heart cells have been detected by similar techniques*

- *Voltage sensitive DHPRs* (dihydropyridine, hence DHP receptors, also called L-type Ca channels), of which there are many in the T-tubule membrane. Depolarization of these causes a conformational change and promotes linkage to the second type of Ca^{2+} ion (release) channel in the membrane of the terminus of the sarcoplasmic reticulum.
- *Calcium ion release channels*, known as ryanodine receptors (RyR1) (Table 3.2). All ryanodine receptors are activated by an increase in $[Ca^{2+}]_{cyto}$. They are therefore called calcium induced calcium release (CICR) receptors.

The close apposition of DHP and Ry receptors (they can be physically connected by gap junctions) allows effective EC coupling. The result is a flood of Ca^{2+} ions flowing from the lumen of the SR where the Ca^{2+} ion concentration is ~mM and storage in particles is aided by Ca^{2+}-ATPase and calsequestrin. The increase in the Ca^{2+} ion concentration in the myoplasm (the cytosol portion of the muscle cell) from the resting value of ~100 nM to 1–10 μM, occasioned by these events is enough to trigger muscle contraction (*vide infra*). At the termination of the depolarisation of the T-tubule, Ca^{2+} ion is sequestered from the myoplasm back to the lumen of the SR using a Ca^{2+}-ATPase pump that is the dominant membrane protein in the SR. Removal of Ca^{2+} ions from the muscle system restores the relaxed state. If successive action potentials release more Ca^{2+} ions from the SR before all the Ca^{2+} has been pumped back, then the cytosol $[Ca^{2+}]$ remains elevated and contraction of the muscle is maintained. This condition is called *tetanus*.

Defective Muscle Can Still Be Effective

Swordfish diving in cold ocean water can maintain cranial temperatures up to 14 °C above that of the surrounding water. Animals of this kind employ heater organs as one source of body heat. The heater organs are simply modified muscle that lacks myosin ATPase activity and contractile elements. They retain, however, an extensive network of SR and T-tubules close to a large volume of mitochondria and use excitation from axons and rapid Ca^{2+} ion release and uptake. Ca^{2+} ion pumping by the SR Ca^{2+}ATPase is a major source of heat, which results from dissociation of ATP provided by mitochondria.

We have previously encountered a modified muscle that could not contract but was able to generate crippling voltages (Figure 4.9).

The restoration of the muscle membrane potential to its resting value is effected by the use of chloride channels located on the T-tubule and muscle surface. The flow of Cl⁻ ions contributes significantly to the repolarization of the muscle cell membrane.

Startling Story of Fainting Goats

The voltage gated chloride channels in skeletal muscle are very important. These channels are mainly responsible for the resting potential of the muscle cell membrane and their role is the same as that of K channels in neurons and most other cells.

There is a breed of goat that originated in Tennessee in the 1880s that tends to develop leg rigidity and fall over when startled. Fortunately, the animals recover completely within about a minute. They are called myotonic goats. Indeed there is an International Fainting Goat Association dedicated to the preservation of the breed! Their unique behaviour was shown to be due to an abnormal muscle Cl⁻ ion channel. This was the first time that a channel malfunction was linked to a clinical disorder. A small increase in extracellular $[K^+]$ in the T-tubules accompanies muscle relaxation. This normally is offset by the dominant chloride conductance. In myotonic muscle however, there is a diminished ability of the chloride channel to open. The rise in extracellular potassium ion will now lead to a small residual membrane depolarization after action potential termination. A significant number of these so-called 'after-depolarizations' can cause spontaneous action potentials (Figure 6.3). These, in addition to the smaller stimulus required to produce an action potential, means that relaxation of the muscle after voluntary contraction is markedly delayed. The study of myotonic goats led to establishing the cause of a similar disorder in mice, miniature Schrauzer dogs and in human patients with so-called *myotonia congenita*. The disorders are due to mutations in the skeletal muscle chloride channel gene, CLC-1, and many point mutations have been identified. For example, the mutation responsible for dysfunction in myotonic goat muscle is A885P in the carboxyl terminus. Myotonia can also be associated with mutations in sodium ion channels (6.6.1).

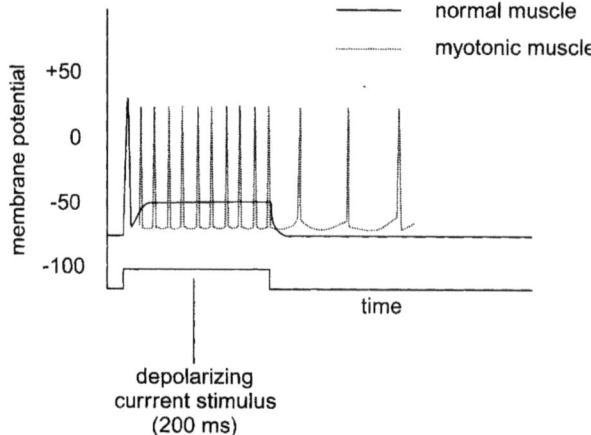

Figure 6.3 *Action potentials associated with normal and myotonic intercostal (between the ribs) muscle fibres, in vitro at 37 °C*

6.2.2 Muscle Contraction

What roles do Ca^{2+} ions have in the final events in muscle contraction? The contraction of muscle fibres is shown conceptually in Figure 6.4.

The major component of the thick muscle filament is myosin. It is an

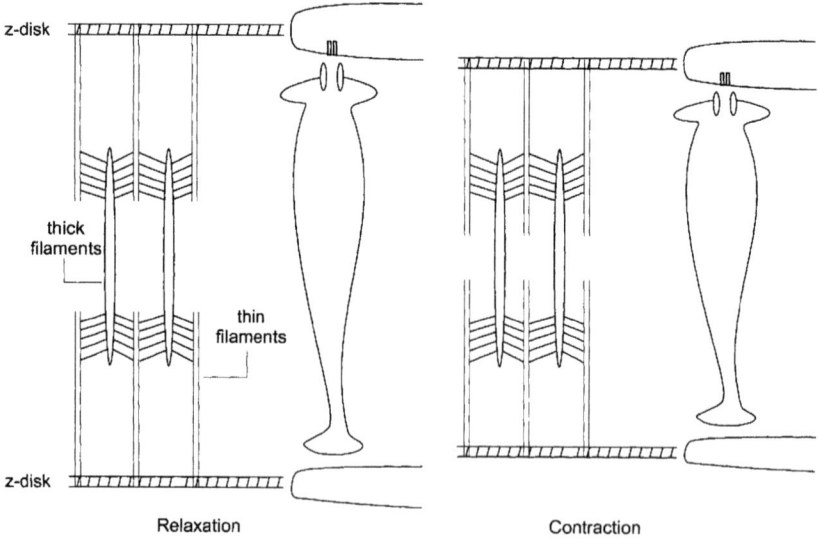

Figure 6.4 *Muscle contraction. A rapid rise in $[Ca^{2+}]_{cyto}$ causes the contractile filaments to shorten*

ATPase that catalyses the conversion of ATP to ADP and P_i, an enzymatic activity that has a major role in the contraction–relaxation cycle. In contrast, the thin filament is made up of a number of proteins, particularly actin, but also tropomyosin and troponin complex (TnC, TnI and TnT). In the relaxed muscle, tropomyosin *blocks* the attachment of the myosin heads to a number of actin units (Figure 6.5). At the higher Ca^{2+} ion concentration generated from the T-tubules and SR, calcium ion binds to the TnC subunit of troponin. The Ca^{2+}–TnC complex (plus TnI and TnT) formed interacts with tropomyosin and a conformational change removes the blockage.

Now the actin and myosin can form a complex and initiate the contraction cycle (Figure 6.5). It is the interaction of the myosin head groups (attached to myosin filaments) with actin filaments that causes muscle contraction. Myosin and actin filaments are polymeric forms of the respective proteins. In the sliding filament 'tight coupling' model (Figure 6.6), which is supported by most researchers, interdigitated sets of thick and thin filaments slide past each other in a continual (cyclic) action by using the crossbridges that project from myosin filaments.

There are four *fundamental* steps in skeletal muscle contraction. However, there are also transient intermediates (involving weak binding of myosin and actin) and other subtleties that are also probably involved. When ATP or ADP and P_i are bound to myosin they are located in a pocket in the S_1 myosin head, which is also the site of myosin/actin interaction. The four basic steps (Figure 6.6) are:

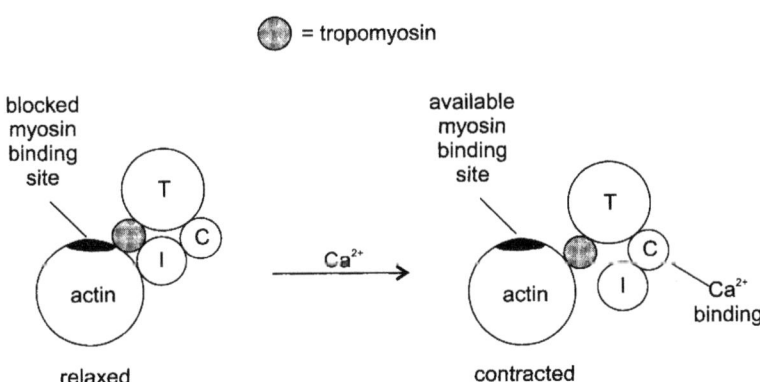

Figure 6.5 *The role of Ca^{2+} ions in removing the blockage to actomyosin formation. The troponin complex of TnT, TnC and TnI (T, C and I) binds to tropomyosin and actin in the absence of Ca^{2+} ions, but not in the presence of Ca^{2+} ions. The COOH domain of TnC is bound to Mg^{2+} ions. The troponin complex is distributed at regular intervals along the entire thin filament*

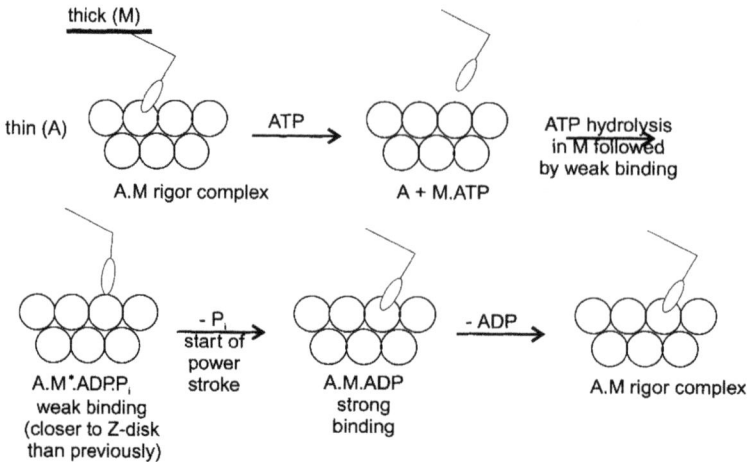

Figure 6.6 *The sliding filament model for muscle contraction*

- The rigor complex A.M involves strong binding of actin (A) to myosin (M) in the absence of ATP. The complex (actomyosin) binds to ATP and in the process the cross bridge is detached and the actin–myosin binding is weakened, giving [A + M.ATP]. It has long been known that the combination of the myosin head with ATP is required to sever the link between myosin and actin. If ATP is unavailable (death) the two filaments remain locked (*rigor mortis*).
- Hydrolysis of the attached ATP catalysed by the myosin provides the chemical energy to produce an energised form of myosin, M^*, that is still associated with ADP and P_i, giving [A + M^*.ADP.P_i].
- The energized form of myosin promotes binding of actin to myosin, to give [A.M.ADP] with the release of P_i.
- A protein conformational change allows the movement of the cross bridge to reform the rigor complex, A.M. This step is called the *power stroke*. However, the thin filament has moved ~10 nm, relative to the thick filament, toward the centre of the sarcomere. ADP is then released and the cycle continues (provided ATP and sufficient Ca^{2+} are present).

In the contracted state, all the possible cross bridges are not required to be linked in order to provide tension.

Tongue Power as Employed by the Chameleon

Z-disks, which are normally solid, limit the extent of muscle contraction when the end of the thick filament crashes into them (Figure 6.4).

The retractor muscle in the tongue of the chameleon has perforated Z-disks that let the thick filament pass through, thus allowing supercontraction. This, plus a large overlap between thick and thin filaments (and great force) enables the chameleon to extend its tongue to more than twice its body length. It can therefore use its tongue to catch even large prey some distance from its mouth.

The general role for Ca^{2+} ions in muscle contraction is illustrated by the simultaneous recording of the membrane potential, the SR calcium ion concentration and muscle tension in a single fibre of the barnacle *Balanus nubilus* Darwin, which is used because of its large size and sturdiness (Figure 6.7).

The $[Ca^{2+}]$ is observed to increase rapidly (B) following membrane depolarization (A) and reaches a maximum soon after the stimulus pulse (D) ceases. It then falls exponentially. The resulting muscle tension change (C) develops more slowly and is controlled, but obviously not completely,

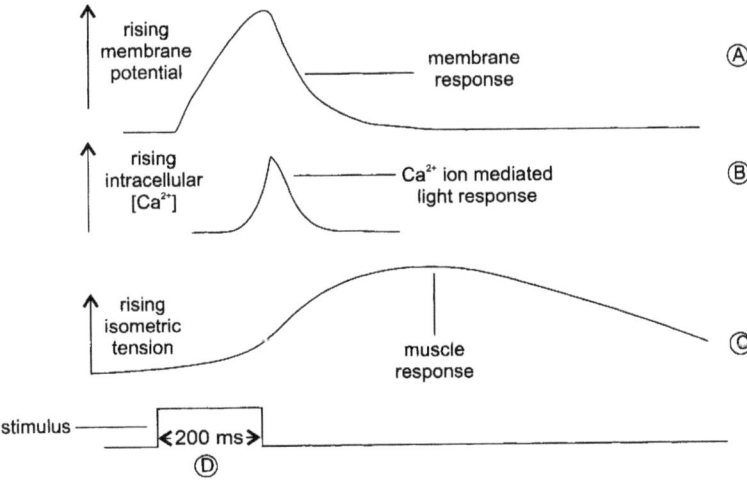

Figure 6.7 *Linking of membrane potential (using an intracellular recording electrode), intracellular [Ca²⁺] (using aquorin or fura-2) and muscle tension in barnacle muscle. A single depolarization pulse is applied to the fibre membrane. The isometric tension is that developed while the muscle is prevented from contracting*

by Ca^{2+} ions. As the stimulus strength increases, the rate and magnitude of the rise of both the $[Ca^{2+}]$ and muscle tension increases (not shown). Another demonstration of the necessity of calcium ions is the use of caged Ca^{2+} (1.5.4). If Ca^{2+} ions are released in the vicinity of a frog myofibril, a rapid rise in tension is observed.

If extracellular $[Ca^{2+}]$ falls to ~40% of the normal value, *hypocalcemic tetany* occurs involving involuntary tetanic contraction of skeletal muscles (muscle cramps). This is because a low external calcium concentration leads to increased opening of Na^+ ion channels in excitable plasma membranes, depolarization and spontaneous firing of action potentials. As a consequence there is increased muscle contraction.

6.2.3 Summary of the Events Between Stimulation and Muscle Contraction

At this point it is worth summarizing the sequence of events that occur between sensation and skeletal muscle fibre contraction.

- The brain receives a signal from a sensory receptor (Chapter 7) and sends a signal along sensory neurons to the posterior roots of the spinal cord, and from there through interneurons to the ventral roots. An action potential is then initiated in the spinal motor neuron proceeding from the cell body along an axon to the terminal (Figure 4.2).
- Depolarization of nerve endings activates voltage gated calcium channels in the nerve endings.
- Calcium ions enter the nerve endings and trigger release of ACh, which diffuses from the axon terminals to the motor end plates in the muscle fibres.
- ACh binds to AChRs triggering the opening of cation channels.
- Na^+ ions move into, and K^+ ions move out of the cell resulting in depolarization (end plate potential).
- The local current depolarizes the adjacent plasma membrane to its threshold potential and an action potential is propagated over the surface of the muscle and into the fibre along transverse tubules causing the depolarization of DHPRs.
- Calcium ions are released from the sarcoplasmic reticulum and initiate contraction by allowing the formation of actomyosin.
- When $[Ca^{2+}]_i$ decreases and calcium ions are transported into the SR by Ca-ATPase, the muscle fibre relaxes.

6.3 MUSCLE FUNCTIONS

Muscles are used in a variety of ways and we will categorize a few.

6.3.1 Posture and Movement of Bodies

Movement using muscles is an obvious function. It appears in a variety of guises with mechanisms that are common to all animals. Examples include man walking, frogs jumping, fish swimming and escape mechanisms.

The Escape Mechanism of Jellyfish Differs From That of Higher Animals

An escape response is frequently mediated by an array of giant neurons synapsing with the appropriate muscles. Examples include the squid *Loligo* (Figure 4.11), the cockroach *Periplaneta* (Figure 7.3), the crayfish *Procambarus* and the jellyfish *Aglantha digitale*.

The *Aglantha digitale*, a member of a phylum (Cnidaria) of aquatic invertebrates, is bell shaped and up to 30 mm in height. It swims by jet propulsion using contracting muscles that constrict the walls of the bell and expel water through the membrane covering the base. It has two swim modes. The usual one (slow) is generated endogenously and is based on rhythmic weak contractions. During a fast (escape) swim initiated by a predator (or by tugging on a tentacle) a violent contraction propels the animal about five body lengths at a peak speed of 0.3 m s^{-1}.

Both contractions arise from giant motor neurons that make multiple direct synaptic contacts with the striated muscle sheet that comprises the inner surface of the body wall. The axons use Ca^{2+} ion generated action potentials for slow swimming and Na$^+$ ion dependent action potentials for fast swimming. This simple device contrasts with higher animals that use groups of structurally different muscle fibres or complex innervation to achieve these different behaviours (6.3.2).

6.3.2 Production of Sound

Male *Drosophila* generate an acoustic signal by vibrating their wings (its 'love song'). The sound is species specific and has a role in species recognition as well as in courtship. A noisy muscle is also important in the mating rituals of other animals. There are three muscle fibre types found in the male toadfish (*Opsanus tau*) that can be relatively easily dissected since they are anatomically segregated. The slow tonic red muscle, with a relatively high myoglobin content, is used for steady swimming. The fast twitch white muscle is used for rapid movement. The superfast swimbladder (sonic) muscle that encircles the gas-filled swimbladder achieves a relaxation – contraction frequency of around 200 cycles s^{-1} (Hz). This is a record rate for a vertebrate and is responsible for the 'boatwhistle' mating call of the creature. The sound can be repeated every five seconds or so, for many hours! The variation of twitch tension and of [Ca^{2+}] with time for each fibre type is shown in Figure 6.8.

It can be seen that the free Ca^{2+} ion concentration parallels the twitch duration. The rapid increase in [Ca^{2+}] required to activate troponin followed by the rapid decrease is necessary for rapid muscle contraction.

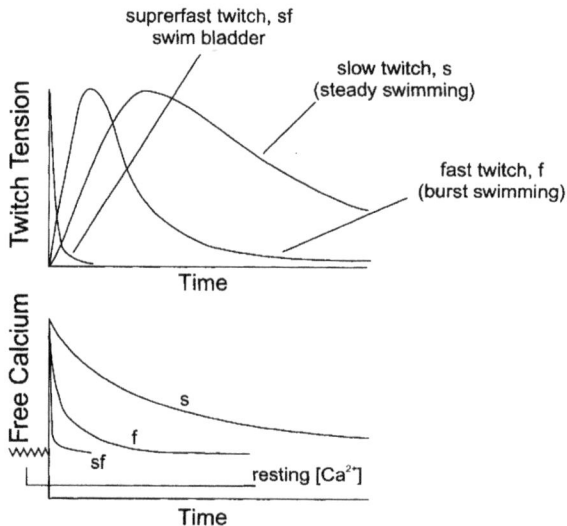

Figure 6.8 *Twitch tensions and Ca^{2+} ion concentrations in toadfish fibres at 16 °C. The values are normalized to their maximum values for all fibre types. Twitch tension half-widths are approximately 500 ms (s), 180 ms (f) and 10 ms (sf). These are accompanied by myoplastic free Ca^{2+} half-widths and concentrations respectively of approximately 110 ms and 7 μM (s), 20 ms and 11 μM (f) and 3 ms and 51 μM (sf). The magnitude of [Ca^{2+}] and the short tension for sf are the largest and fastest recorded for any fibre type. The half-width is the width between ascent and descent of a trace at half height. The twitch tension is the force produced by the muscle as it is kept from shortening (isometric contraction). The Ca^{2+} ion concentration is measured by using furaptra, a fluorescent indicator that is injected by pressure into a single fibre*
Reprinted from L. C. Rome *et al.*, *PNAS*, 1996, **93**, 8095–8100, copyright 1996 National Academy of Sciences, USA.

The transient rapid [Ca^{2+}] increase may arise because of a larger SR and a higher density of calcium channels in the superfast muscle type. Two other requirements for a superfast type fibre would be a rapid rate of dissociation for the Ca^{2+}–troponin complex and faster cross bridge detachment, both of which have been confirmed recently. Fast relaxation is aided by the presence of parvalbumin, a strong calcium ion binder that is not found in humans, which facilitates the transfer of Ca^{2+} ions between myofibril and SR.

Method in the Madness of the Squirrel

The fibres of the shaker muscle of the rattlesnake (not used in their mating ritual!) have characteristics similar to those of the toadfish swimbladder. The rattle of the tail is relatively slow (~60 clicks s^{-1}) at 10 °C, but can be as high as 200 s^{-1} at 35 °C. The California ground squirrel uses this information to determine whether the northern Pacific rattler is cold (and therefore more sluggish) or whether it has warmed itself in the sun and is thus more likely to be quicker and deadlier. The squirrel kicks sand in the face of the snake to induce it to shake its rattle and produce the desired frequency information! This safety device has probably been in place for ten million years.

6.3.3 Insect Flight

Insects communicate by means of the distinctive sounds of their wing-beats. Insects that have a low wingbeat frequency of about 4–25 beats s^{-1}, as in butterflies, cockroaches and locusts, are using wing movements that are in unison with the muscle action potentials that generate them (Figure 6.9). This is a synchronous muscle action. However, when the wingbeat frequency is higher, as it is with most insects, then the activating neuron input and muscular action potentials are too slow for the high frequency of oscillating contractions (Figure 6.9).

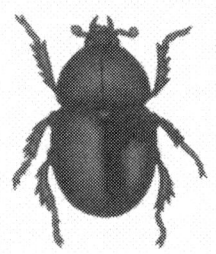

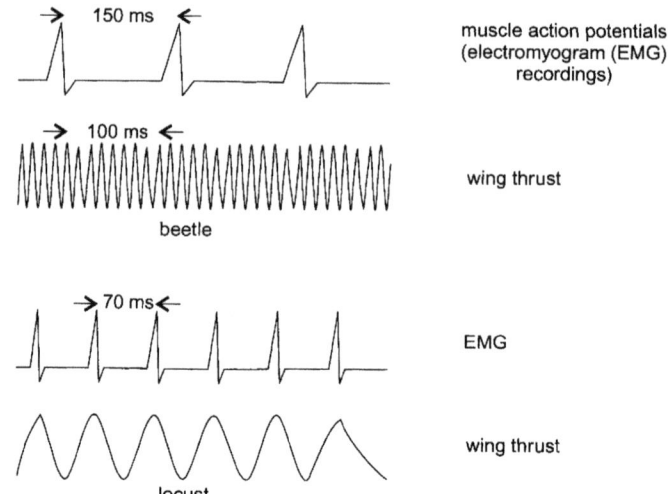

Figure 6.9 *Muscle action potentials and associated wing thrusts for the muscles of the locust Schistocerca americana and the beetle Cotinus mutabilis. EMGs were recorded via silver wires that were implanted in the wing muscles and a force transducer recorded the thrust of the wing beats during tethered flight. The wing-stroke frequency of the locust muscle was 16 s⁻¹ and that of the beetle muscle was 76 s⁻¹*

Neuron input is still required to start muscle contraction, but once initiated the (so-called asynchronous or fibrillar) muscle continues to contract since it is attached to the elastic thorax, and the muscle and thorax comprise a mechanical oscillator. Only insects, beetles, bees and flies for example, use this mechanism to attain wingbeats of around 100 beats s⁻¹. It should be noted however, that synchronous muscles are also known with these higher frequencies (see rattlesnake and toadfish, above). Muscle activation in both synchronous and asynchronous muscles, shown in Figure 6.9, is the result of Ca^{2+} ion release from the SR.

6.4 CARDIAC MUSCLE

6.4.1 Comparison with Skeletal Muscle

When one is confronted with the fact that the human heart contracts and relaxes more than three billion times over a normal lifetime, the magnificence of the cardiac muscle can be appreciated. Although cardiac muscle resembles skeletal muscle in many respects (Table 6.1), there are important differences in the events that precede contraction. Nerve impulses initiate contraction in skeletal muscle whereas cardiac muscle has intrinsic rhythmicity, that is it can contract rhythmically without any external stimulus. During heart muscle relaxation the chambers of the heart fill with blood. During contraction, blood is propelled from the heart.

An Isolated Frog's Heart Takes a Licking and Keeps on Ticking

A frog heart that has been removed from the creature continues to beat for hours. The times are shorter for mammalian hearts. The heart in a developing chick embryo beats even before any nerves are attached to it. However a saline solution containing Ca^{2+} ions is essential for the continuing beat of an isolated heart (first shown serendipitously in 1883 using a frog heart and Central London tap water), whereas skeletal muscle can continue to contract for some time without an extracellular source of Ca^{2+} ions. It is apparent that extracellular Ca^{2+} ions are even more important in cardiac than in skeletal muscle. In fact, most of the Ca^{2+} ions in amphibian (*e.g.* frog) heart cells enter through the surface membrane during depolarization.

Adult mammalian heart function still uses the amplification of a $[Ca^{2+}]_i$ signal by CICRs employing coupling of DHPR/RyR channels. In contrast to skeletal muscles these channels (6.2.1) are probably not connected but are separated by a cleft of about 10 nm. Decreased ability of DHPR channels to activate adjacent RyR channels may explain various cardiac dysfunctions. The trigger signal for relaxation, as with skeletal muscle, is the return of the membrane potential to its resting value. Ca^{2+} ions exit the myocytes and are pumped back into the SR primarily *via* the Ca^{2+}/Na^+ exchanger. The Ca^{2+}-ATPase pump and an ATP promoted Ca^{2+}/K^+ exchanger are involved to a lesser extent than in skeletal muscle.

6.4.2 Generation of Action Potential

The various steps in the generation of a cardiac muscle action potential are shown in Figure 6.10. The action potential generated in cardiac muscle is of much longer duration than that in skeletal muscle, largely due to the role of Ca^{2+} ions (phase 2). Phases 1 and 2 have no real counterpart in skeletal muscle.

Because an action potential is concluded long before contraction in skeletal muscle, repetitive stimulation and tetanic contraction is possible.

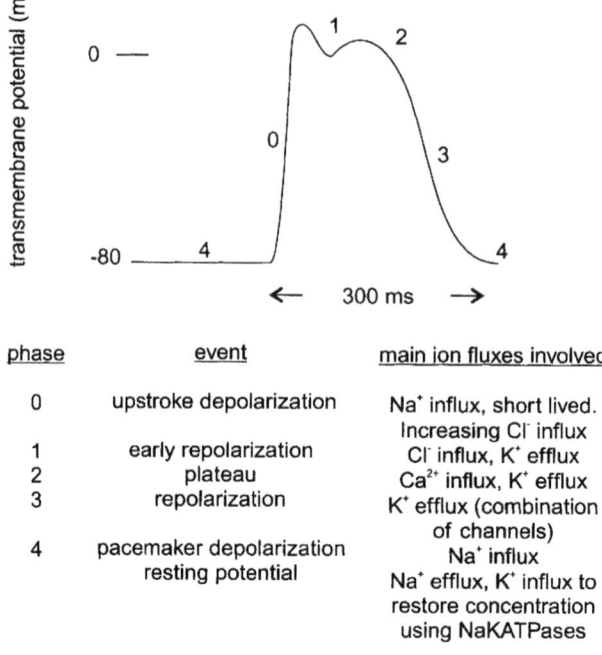

phase	event	main ion fluxes involved
0	upstroke depolarization	Na⁺ influx, short lived. Increasing Cl⁻ influx
1	early repolarization	Cl⁻ influx, K⁺ efflux
2	plateau	Ca²⁺ influx, K⁺ efflux
3	repolarization	K⁺ efflux (combination of channels)
4	pacemaker depolarization	Na⁺ influx
	resting potential	Na⁺ efflux, K⁺ influx to restore concentration using NaKATPases

Figure 6.10 *Cardiac action potential (Purkinje fibre). At least ten potassium channel genes encode for the plateau and repolarization phases*

With cardiac muscle however, the membrane remains in a refractory state (4.4.2) (for about 250 ms) until the heart has returned to a relaxed state. In the refractory state, the muscle does *not* respond to a stimulus of any magnitude.

Cardiac impulse generation and conduction are shown in Figure 6.11. All cardiac muscle cells are electrically interconnected. Once initiated in pacemaker cells (see below) the electrical impulse (action potential) spreads throughout the heart muscle, just as in nerve signalling. In fact, pacemaker cells *are* neurons in many invertebrate hearts.

The electric impulse for contraction originates in pacemaker cells of the SA node. These set the tempo of the heartbeat. The resting potential of the SA node is not steady, but undergoes slow depolarization (pacemaker potential). This arises from a movement of Na⁺ ions through the voltage gated Na channels that are opened by the repolarizing phase of the preceding action potential. (This is quite different from voltage gated Na channels in nerves that are opened by depolarization). When the pacemaker potential is brought to threshold, the resulting action potential, generated 60–100 times a minute, spreads rapidly through the muscles of the two atria. After a slight delay it proceeds to the muscles of the

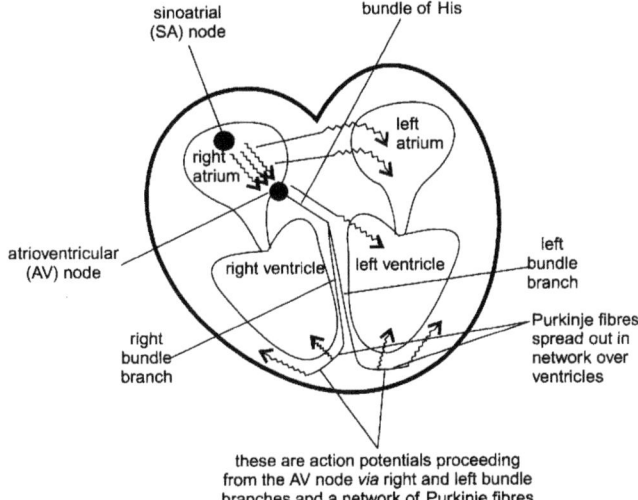

Figure 6.11 *Cardiac impulse generation and conduction in mammals. The action potentials reach the overlying muscle forcing contractions first at the top of the heart and then at the bottom. These contractions are used to squeeze oxygen-depleted venous blood out of the right atrium into the right ventricle and thence into pulmonary circulation. Oxygen is resupplied to the blood that is delivered to the left side of the heart. The muscle pumps this oxygen rich blood from the left atrium to the left ventricle and then out to the aorta and body. The right and left sides are separated. Other vertebrates can have different numbers of atria and ventricles*

ventricles *via* the AV bundle (which is a partition between atria and ventricles) and finally reaches the His–Purkinje fibre system, which conducts impulses very rapidly. Action potentials observed in different parts of the heart vary quite significantly. Our picture (Figure 6.10) illustrates the action potential in Purkinje fibres, since this is the most studied tissue. Following the action potential peak, the membrane repolarizes and gradual depolarization recommences leading to spontaneous, rhythmical self-excitation.

6.4.3 Electrocardiogram (ECG)

An easily recognized manifestation of action potentials in cardiac muscle is the electrocardiogram (ECG). The ECG provides a record of the electrical events occurring within the heart by using high conductance jelly and electrodes placed on the skin surface. These pick up weak voltage changes arising from current flowing *outside* cells during the passage of the action potential. It is the sum of the electrical activity in various cells of a heart. The relationship between the electrocardiogram and ventricular and atrial action potentials is shown in Figure 6.12.

The measurement of the QT interval and the duration of the QRS complex allows for clinical assessment of the effects of drugs and of diseases on the properties of the ion channels responsible for ventricular depolarization and repolarization.

6.4.4 Cardiac Arrhythmias

If various sets of cardiac muscles are not in rhythm, then abnormalities in heart rhythm such as increases (tachycardia) and decreases (bradycardia) in rate can arise, resulting in irregular pumping of blood. Antiarrhythmic drugs are used to correct abnormal heart rhythms by *blocking* various

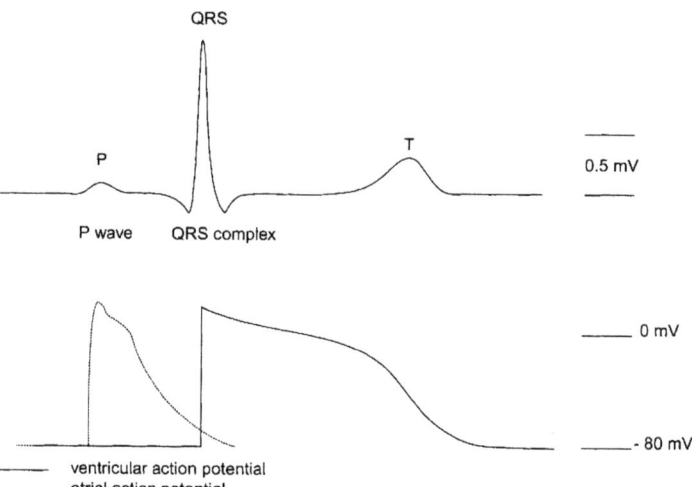

Figure 6.12 *Electrocardiogram (upper) and related cardiac action potentials (lower). In the ECG, the small P wave corresponds to atrial depolarization and contraction and represents the signals passing from SA to AV nodes (Figure 6.11). The largest peak (QRS) is the extracellular record of the upstrokes of the ventricular action potentials (phase 0, Figure 6.10), with signals passing rapidly along the bundles of His, bundle branches and Purkinje fibres. The repolarization and relaxation phase of the ventricular action potential appears some 300 ms later as the slower and smaller T peak. The ECG is a much smaller signal than the action potentials (compare the scales) and because of this peaks are recorded only when the membrane potential is changing, i.e. the ECG is flat in the plateau phases (PR and ST intervals) of the action potential*
     Reprinted with permission from Blackwell Publishing from H. Brown and R. Kozlowski, *Physiology and Pharmacology of the Heart*, Blackwell Science, Oxford, 1997.

channels and slowing the response of heart muscles to normal stimulation. Examples are shown in Table 6.2.

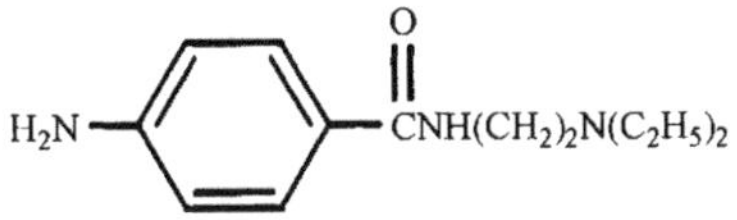

(1)

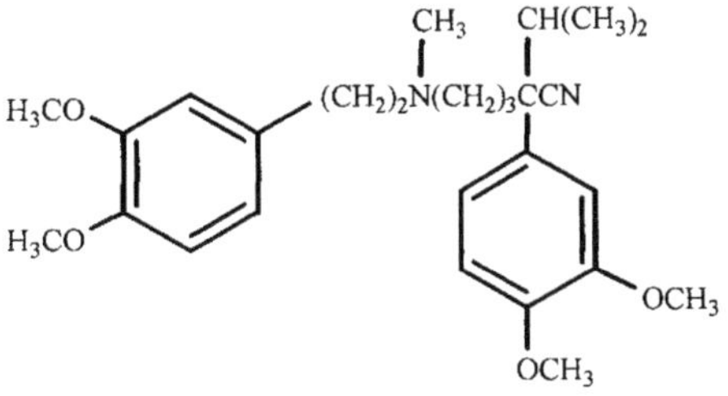

(2)

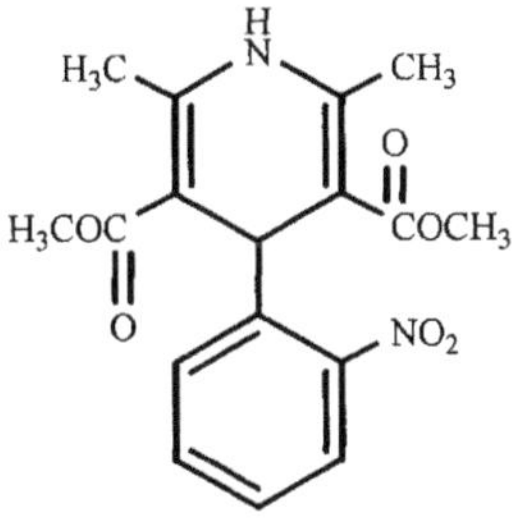

(3)

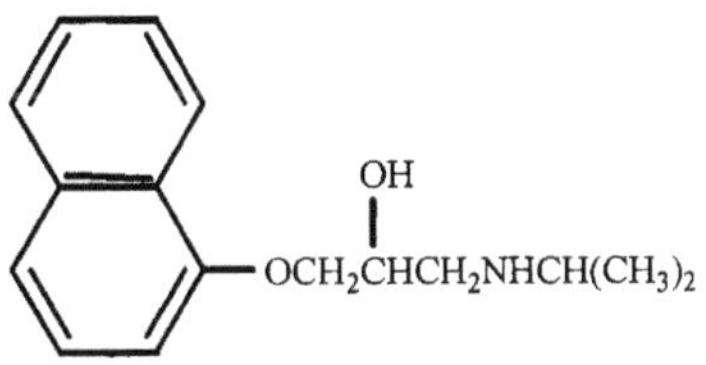

(4)

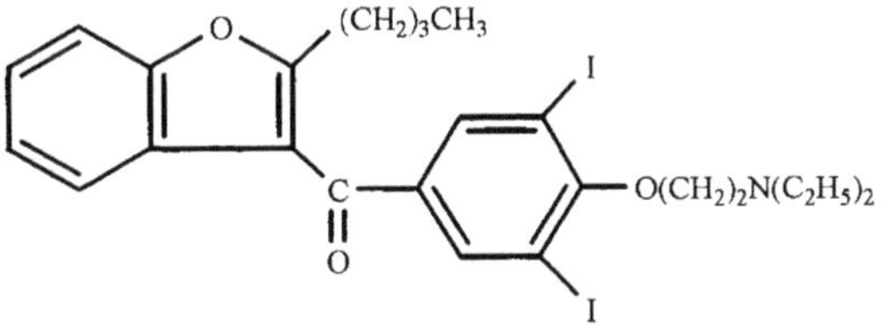

(5)

These drugs are often used in life-threatening situations. They can, however, have serious side effects and clinical trials have shown that persons taking these drugs (except β-blockers) do not live longer than those who are not! Cardiac glycosides have been prescribed for congestive heart failure and other heart conditions. The most famous example has been used for centuries (digitalis).

Table 6.2 *Antiarrythmic drug actions on various channels in cardiac muscles*

Channel	Phase of Action Potential (Figure 6.10)	Examples of Drugs
Na	0	Procainamide (1) (action is identical to that of a local anesthetic).
Ca	1–2	Verapamil (2) and Nifedipine (3) are antagonists of Ca L-channels. Propranolol (4) is a β-adrenergic blocking agent. The (–) enantiomer is 10^2 times more effective than the (+).
K	2–3	Amiodarone (5) (prolongs action potential).

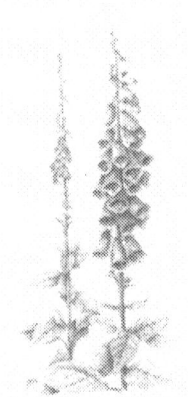

Poisonous Plant Provides Positive Effect

Digitalis is a mixture of steroidal glycosides found in the leaves of red foxglove (*Digitalis purpurea*). The plant is poisonous to animals, but extracts of foxglove were used for the treatment of heart failure as early as 1785. Digitoxin (6) (also called digitalin) and digoxin (7) are components of digitalis. They are widely used to treat heart failure and arrhythmias. They act by binding to the external exposed portion of the Na^+/K^+ ATPase pump.

(6) X = H; (7) X = OH

This reduces the exit of Na^+ ions and therefore increases $[Na^+]_i$. Now sodium ions must be pumped out by some other means, and the Na^+/Ca^{2+} exchanger is used for this purpose. However, as a result, Ca^{2+} ions enter the cells, including those of the SR. This extra $[Ca^{2+}]$ increases the force of cardiac contraction (*positive ionotropic effect*) and is of benefit to a failing heart. The effect is a favourite of murder mystery novelists since an overdose of digitalis is lethal due to precipitation of ventricular fibrillation, which is the major cause of sudden cardiac death. Irregular cardiac electrical waves and a heart rate exceeding 500 excitations per minute causes inadequate pumping of blood. Oubain (8), a poison (used on arrowheads) from the East African Ouabio tree, (and now found edogenously in the adrenal cortex) serves a similar purpose by inhibiting Na^+/K^+ ATPase. However, the greater number of OH groups in oubain reduces its membrane permeability and thus it requires parenteral administration. All three are used as experimental tools to block the Na^+ pump.

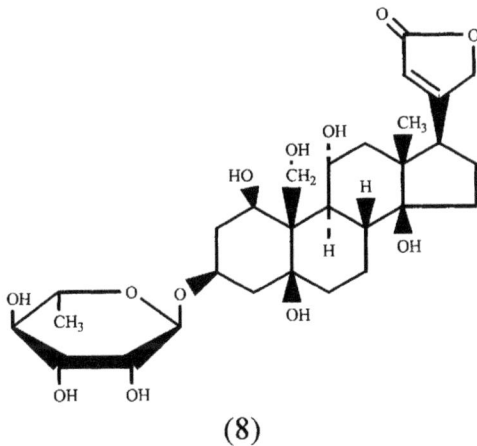

(8)

Intracellular acidosis, a fall of pH_i resulting from, for example, myocardial ischcmia, introduccs dangerous conditions in the myocardium (see Figure 3.9). The Na^+/H^+ antiporter and the Na^+/HCO_3^- symporter (Figure 3.9) remove H^+ ions, which is desirable, but at the expense of accumulating Na^+ ions in the cell. Acidosis can also be accompanied by inhibition of Na^+/K^+ATPase and thus retard the flow of Na^+ ions out of the cell, again leading to the accumulation of sodium ions in the cell. Removal of Na^+ ions by the Na^+/Ca^{2+} exchanger is at the expense of introducing excess Ca^{2+} ions (Ca^{2+} overload) and cardiac disturbances (see digitoxin). One treatment of acidosis is to use *amiloride* (3.6.1), a diuretic, which inhibits

Na^+/H^+ exchange by competing with Na^+ for the extracellular Na^+ binding site. The increases in $[Na^+]_i$ and in $[Ca^{2+}]_i$ are thus prevented. Amiloride is an effective drug that is used in postischemic recovery.

Mutations in genes that encode myosin, troponin T, troponin I and other proteins involved in contraction can cause abnormalities that result in muscle malfunction. Abnormally thick heart walls, for example, can cause the heart to pump strongly, but to not relax very well. In other cases the heart 'flutters' and does not pump blood ('fibrillation').

Heart Failure Can Result From a Mutated Cardiac TnT

A deletion mutation ΔK210 in cardiac troponin T (cTnT) has recently been found to cause familial dilated cardiomyopathy (DCM) a disease with a high incidence of death in young adults. The free Ca^{2+} ion concentrations required for force generation in rabbit cardiac muscle fibres containing mutant cTnTs were higher than in wild type. The primary mechanism for pathogenesis of DCM associated with mutation ΔK210 is a deficiency of force generation by the sarcomere in cardiac muscle.

6.4.5 Nervous Control of Heartbeat

The depolarization of pacemaker cells mentioned earlier (6.4.2) is a determining factor in the heart rate and the force of contraction. Cardiac muscle has a built-in frequency of 100 beats min^{-1}. *Sympathetic* nerve impulses (Figure 6.13) increase the rate of depolarization and thus the heartbeat. Activity in the cardiac sympathetic branch is raised by anger or fear (or vigorous exercise) and the nerve terminals release the neurotransmitter norepinephrine (9) (noradrenalin), the precursor of epinephrine, from the adrenal gland into cardiac muscle cells. Epinephrine (adrenaline) (10) pours into the blood stream (it can increase 10-fold from 1–3 nM in resting plasma) in response to stress and prepares the body for action. This happens in the following manner.

(9)   (10)

Epinephrine binds to the β_1-adrenergic receptor on the surface of the cardiac cell. This activates a G-protein that promotes adenylate cyclase conversion of ATP to cAMP. Kinase A is activated by cAMP and can then phosphorylate, amongst other things, the β-subunit of an L-type Ca^{2+} channel. This leads to increased opening of the channel and enhanced Ca^{2+} ion entry. This speeds up the pacemaker cell depolarization and the calcium ion based upstroke of the action potential. The overall result is an increased rate and force of cardiac muscle contraction. As a consequence, the heart beats faster and blood is circulated more rapidly allowing the epinephrine-aided 'fight or flight' response to danger.

The sympathetic nervous system thus releases epinephrine on to cells of the SA and the AV nodes (Figure 6.11) and speeds up heartbeats. In addition however, the autonomic involuntary system (that is not under

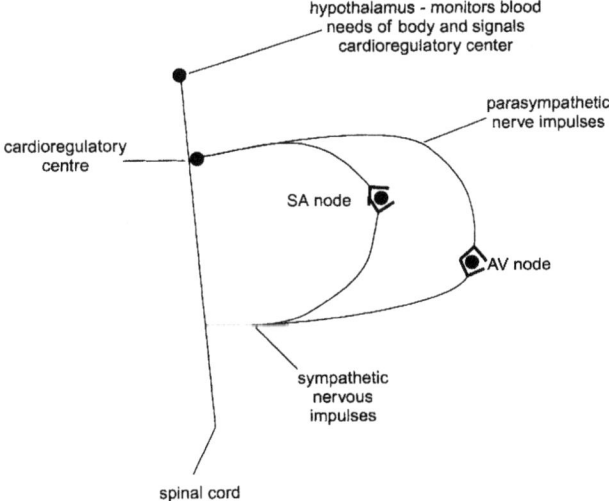

Figure 6.13 *Autonomic nervous system connection to the heart. The autonomic nervous system is responsible for the control of involuntary body functions such as blood pressure, sweating and digestion as well as for the heart rate*

conscious control) also comprises parasympathetic nerves. These synapse with the same cardiac nodes (Figure 6.13), but are inhibitory and thus have a quiescent function in slowing the heart rate to 60–80 beats m⁻¹. How does this occur? ACh combines with a *m*AChR in atrial heart muscle and activates the G-protein, producing the fragments α-GTP and $\beta\gamma$ dimer. The latter binds to a K_{ir} channel, opens it, and the resulting flow of K^+ ions decreases the rate of depolarization of pacemaker cells and the duration and frequency of the cardiac action potential, thus slowing the heartbeat.

6.5 SMOOTH MUSCLE

A wide variety of stimuli (Table 6.1) initiate contraction of smooth muscles, depending on tissue type, but all operate by changing the concentration of Ca^{2+} ion in the myoplasm. Again, Ca^{2+} ions are both 'global' (*via* voltage gated calcium channels in plasma membranes) and 'local' (*via* the SR). Electrical stimulation of some smooth muscles can spread contraction to neighbouring cells through gap junctions. There are two kinds of smooth muscle. Single unit smooth muscle (in the walls of vertebrate visceral organs) can be synapsed by neurons, but neuronal input is not required for contraction. In multiunit smooth muscle (*e.g.* in the iris of the eye) each cell acts independently and contracts only when a synaptic input is received from a neuron.

Many smooth muscles generate action potentials or small depolarizations (<10–20 mV). These arise from neurotransmitter release from nerve endings. The neurotransmitters bind to receptors that are widely distributed over the entire cell surface rather than being more isolated and structured as for example, with the skeletal neuromuscular junction. Neurotransmitters such as ACh, which are so important in skeletal muscle, also act on smooth muscle, but on different receptors (*m*AChRs).

The Eyes Have It

Acetylcholine acts on muscarinic acetylcholine receptors (*m*AChR) to initiate contractions in smooth and cardiac muscles. Atropine (11) inhibits ACh binding to these receptors (a competitive antagonist) and thus promotes muscle relaxation. It is therefore used medically to decrease stomach spasms, the secretion of digestive juices, to promote pupil dilation and at higher doses to increase heart rates. The deadly alkaloid atropine is present in *Atropa belladonna* (deadly nightshade), named after Atropos (curtailed life in Greek mythology) and the fact that it was once used by Italian women to dilate the pupils of their eyes to enhance their beauty! It is still used by eye doctors for pupil dilation. Hyoscine (12) (or scopolamine) has a structure and behaviour similar to that of atropine. It is obtained from the plant species *solanacaea*, and is used medically for anti-motion sickness and as a truth serum (as in *The Guns of Navarone*). The structures of both alkaloids have *general* similarities to that of acetylcholine (Chapter 3 (5)), which means that they can bind to the ACh receptor and block ACh binding without turning on the receptor. Both (11) and (12) occur naturally as single enantiomers, but racemize easily in solution.

(11)   (12)

Hormones and other agonists are, in general, more important ligands than ACh for smooth muscle receptors. The receptor is not coupled directly to a channel and the transmission of signals is slow, lasting for seconds, compared to the corresponding time (ms) in skeletal muscle. A G-protein linked to the DAG/IP$_3$ system (Figure 3.6) is used a great deal in smooth muscles. In all cases voltage gated calcium channels (rather than sodium channels) are activated. Influx of Ca^{2+} ions to the cell causes further depolarization, [Ca^{2+}] increases in the cytosol (from about 0.1 μM to 0.5 μM) and contraction results. Smooth muscles may also use the release of Ca^{2+} ions from the SR, which is the most important storage organelle for this cation.

In large multicellular organisms, the circulatory system (cardiovascular system) comprises a pump (heart) that pushes blood through a set of interconnected blood vessels (arteries, veins and capillaries). Distortions of the walls of the arteries by thickening or hardening (atheroschlerosis) cause decreased blood circulation and are the major causes of heart diseases and attacks. Kawasaki disease is a sudden illness mainly in young children that is accompanied by a fever. There are a number of other symptoms including changes in heart rhythm. The cause is unknown, but in a limited study it was shown that four of five children with the disease who had lingering aneurysms were also found to have calcium deposits at the site of the aneurysm.

Red Wine is Good for the Heart

Endothelin-1 is a 21 amino acid peptide that is a highly potent vasoconstrictor that can cause coronary atheroschlerosis. It binds predominantly to receptors on the smooth muscle cells of blood vessels. Certain components of red wine have been found to reduce the formation of endothelin-1. This could account in part for the low incidence of cardiovascular problems in the red wine drinking regions of southern France in spite of a diet high in saturated fats. Interestingly, three cardiotoxic peptides called sarafotoxins have been isolated from the venom of the burrowing asp. The toxins are also 21 amino acid peptides with striking sequence homologies to endothelin-1 and they are known to cause bradycardia followed by coronary vasoconstriction.

The greater importance of extracellular Ca^{2+} ions in the initiation of cardiac and vascular smooth muscle contraction is shown by the effect of three classes of blockers that act on L-type calcium channels. Benzothiazepines (*e.g.* diltiazem (13)), dihydropyridines (*e.g.* nifedipine (3)) and phenylalkylamines (*e.g.* DL-verapamil (2)) bind to separate but adjacent receptor sites. In so doing they retard contraction and aid dilation of veins and blood vessels. Therefore they are used particularly in treating high blood pressure (vasodilation), but also to treat angina pectoris (chest pains arising from a shortage of oxygen to the heart, often due to hardening of the arteries). A number of such calcium antagonists have been shown to be present in the roots of plants used for centuries in Chinese herbal medicines. Their structures are all based on that of a simple isoquinoline, papaverine (14) a smooth muscle relaxant and a constituent of opium. These drugs have little effect on skeletal muscles that rely *mainly* on internal stores as their calcium source. Calcium channel blockers are the most widely prescribed drugs in medicine.

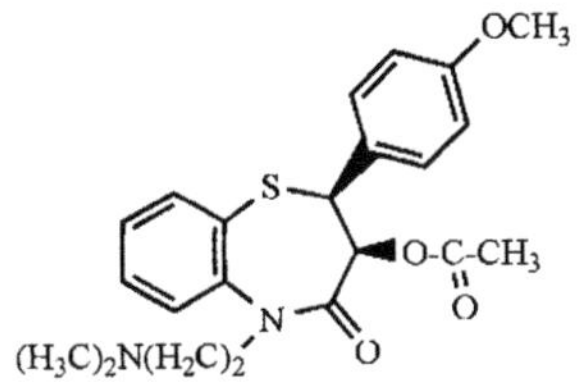

(13) 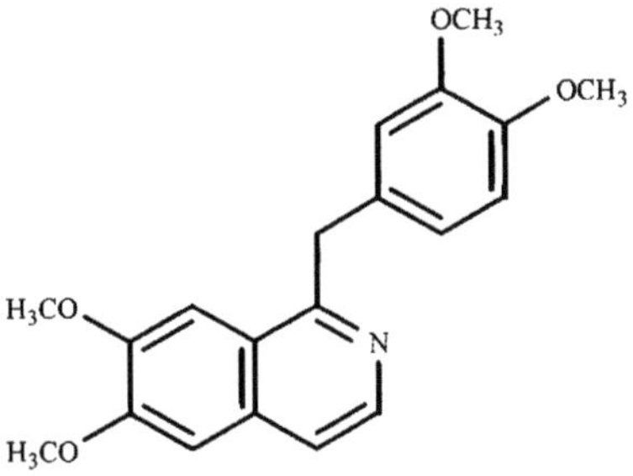   (14)

6.5.1 Mechanisms of Contraction

The contraction of smooth muscles occurs by a radically different mechanism from that used by skeletal and cardiac muscles. No troponin is present in the thin filaments of smooth muscle and thus it is not available to regulate the interaction of myosin with actin as it does in striated muscle. An increase in $[Ca^{2+}]$ initiates polymerization of myosin into the requisite thick filaments that are not normally present (again unlike striated muscle). Calcium ions also play a vital role in one of the major routes of smooth muscle contraction involving the activation of myosin by a phosphorylation pathway. The detailed mechanism is shown in Figure 6.14.

The binding of four Ca^{2+} ions to calmodulin, C, (the number is uncertain *in vivo*) induces a conformational change in the protein $4Ca^{2+}.C$ that exposes hydrophobic sites to which myosin light chain kinase

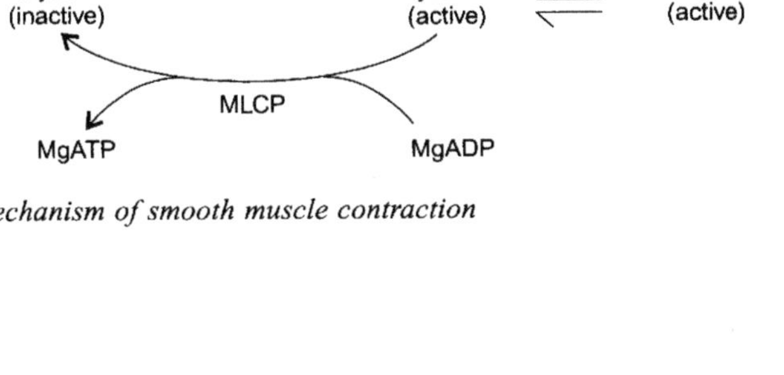

Figure 6.14 *Mechanism of smooth muscle contraction*

(MLCK) binds. The active entity $4Ca^{2+}$.C.MLCK can catalyse the phosphorylation of both of the light chains (small subunits) of myosin by transferring a terminal phosphoryl group from $MgATP^{2-}$ to each ser-19 of the myosin. The (now active) phosphorylated myosin (myosin.P) can interact with actin (dephosphorylated myosin cannot) to form actomyosin.P. In the actomyosin.P complex the cycling of crossbridges from myosin.P along actin filaments constitutes contraction (as with skeletal muscle in the sliding filament mechanism). Relaxation of the muscle is initiated by a fall in $[Ca^{2+}]$ *via* extrusion into the SR, which allows dephosphorylation of myosin.P, catalysed by myosin light chain phosphatase (MLCP). The relative activities of the enzymes MLCK and MLCP will determine whether contraction (MLCK>>MLCP, higher $[Ca^{2+}]$) or relaxation (MLCP>>MLCK, lower $[Ca^{2+}]$) occurs. Other calcium ion dependent (see Catch Muscle) and calcium ion independent mechanisms for EC coupling are also known.

6.5.2 Catch Muscle

Smooth muscles can maintain contraction for long periods of time with much lower expenditure of energy than there would be with skeletal muscles. This ability is used by several molluscs to keep their shells closed for hours or even months (catch muscle). This is important for their protection when the water recedes in an intertidal zone.

Bivalve molluscs, clams and mussels for example, also have a striated adductor muscle, which can contract quickly. Both catch and striated muscles of molluscs and many invertebrates contract by yet another somewhat more direct pathway by the binding of Ca^{2+} ions to myosin light chains, promoting a myosin conformational change that allows it to bind to actin.

Catch Needs Considerable Calcium

The mechanism of catch is still poorly understood. Major factors in promoting this unusual state undoubtedly include:

- Catch muscle contains a high concentration of paramyosin, a muscle protein found in many invertebrates. It has relatively long thick filaments that can develop high tension levels and reduced rates of crossbridge detachment.
- Calcium ion fluxes accompanying catch contraction *in vivo* have been measured using a special non-invasive calcium-selective electrode. Catch response in the anterior byssus retractor muscle of blue mussel (*Mytilus edulus*, a favourite mollusc for study) was generated by carbachol (15) (an acetylcholine (16) analogue). There is a large sustained uptake of Ca^{2+} ions from the surroundings by the muscle (up to 80 pmol $cm^{-2}\,s^{-1}$) followed by a calcium ion efflux that is roughly equal to the influx. The duration of catch is reduced in the presence of 5-hydroxytryptamine (serotonin, Chapter 3 (8)). The mixed nerve to the muscle contains both excitatory and relaxing nerve fibres. It can release acetylcholine (excitatory, a rise in $[Ca^{2+}]_i$ and contraction) or serotonin (leading to activation of adenylate cyclase, cAMP production, a fall in $[Ca^{2+}]_i$ and relaxation).

$$NH_2\text{-}\overset{\overset{\displaystyle O}{\|}}{C}\text{-}O(CH_2)_2N^+(CH_3)_3 \qquad\qquad CH_3\text{-}\overset{\overset{\displaystyle O}{\|}}{C}\text{-}O(CH_2)_2N^+(CH_3)_3$$

(15)   (16)

(15)(16) *Carbachol can replace acetylcholine and is more resistant to hydrolysis. Acting as a muscarinic agonist, it is used in the treatment of glaucoma by constricting the pupil and allowing drainage of aqueous fluid from the anterior chamber and thus lowering intraocular pressure*

6.5.3 Abnormalities in Smooth Muscle

Contractile abnormalities of smooth muscle are often the causes of diseases. Asthma and bronchitis are due to the unnecessary contraction of bronchial smooth muscle. High blood pressure results from the undesirable contraction of vascular smooth muscle. These conditions can arise from the presence of excess excitatory agonists (*e.g.* histamine) or from an enhanced response to normal stimuli. Most drugs in current use (bronchodilators; *e.g.* alkaloids like ephedrine (17), found in some evergreen shrubs and an analogue of epinephrine (10) and vasodilators; *e.g.* nitroglycerin (18)) target the contractile regulatory mechanisms, in particular the level of cytosolic Ca^{2+} ion concentration.

$$H_3C-\underset{\underset{H}{|}}{N}-\underset{CH_3}{\overset{|}{C}}H-CH-\text{(phenyl)}$$
$$OH$$

(17)

CH$_2$ONO$_2$
CHONO$_2$
CH$_2$ONO$_2$

(18)

Ca^{2+} Helps NO Relax Smooth Muscle

Physicians have used nitroglycerin for at least 100 years to alleviate chest pains arising from angina pectoris. With catalysis by a mitochondrial enzyme, the nitroglycerin is converted into the active agent NO. Chemical messengers such as hormones bind to receptors on endothelial cells lining arteries. This binding opens calcium channels and allows an influx of Ca^{2+} ions into the cell. Calcium bound calmodulin (2.5) activates endothelial nitric oxide synthase, which generates NO from L-arginine. Nitric oxide diffuses from the endothelial cells into the vascular smooth muscle cells that underlie them. Here NO activates soluble guanylate cyclase and an increased concentration of cGMP results (Chapter 2, Equation 2). This moves Ca^{2+} ions into storage area and the lower [Ca^{2+}] that results causes relaxation (and dilation) of the blood vessels.

Nasal decongestants inhibit the contractile response by helping generate phosphorylated MLCK, which does not bind strongly to Ca^{2+} ions. Proliferative (rapidly increasing) disorders of vascular smooth muscle can arise from alterations in mediated signal transduction and are of particular importance in coronary artery diseases. Calcium channel blockers reduce the amount of Ca^{2+} ion entering vascular smooth muscle cells. Therefore the smooth muscles contract less strongly, peripheral resistance is lowered and blood pressure falls. Sodium and potassium channels are largely unaffected by the calcium channel blockers.

Women are less susceptible than men to cardiovascular diseases as long as their blood estrogen levels are high (*i.e.* until menopause). Calcium ('local' not 'global') regulated potassium channels (maxi-K or BK, Table 3.1) are key regulators of vascular smooth muscle cells. The major circulating sex hormone in premenstrual women is 17β-estradiol (19). The extracellular binding of 17β-estradiol to the β-subunit of a maxi-K channel (Figure 6.15) leads to more frequent opening of the channel. This, in turn, causes an increase in the efflux of K^+ ions from the muscle cell, and a more negative membrane potential. The negative membrane potential brings about the closing of Ca^{2+} ion channels resulting in less influx of calcium ions and relaxation of muscles in the walls of blood vessels. This rapidly induced dilation of blood vessels allows increased blood flow and lowers blood pressure. This does not happen in skeletal muscles where maxi-K channels contain only α-subunits.

(19)

6.5.4 Non-muscle Cells

Many non-muscle cells also use the actin–myosin contraction cycle employed in smooth muscles. Examples are cytokinesis (Figure 6.16) (the division of cells into two following mitosis) and amoeboid locomotion.

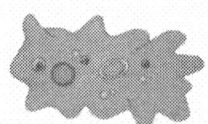

Amoeboid Locomotion

Although the detailed mechanism underlying amoeboid locomotion is uncertain, interconversion of actin filament meshwork (gel) and filament (sol) forms of actin, and conversion of actomyosin into cell movement are

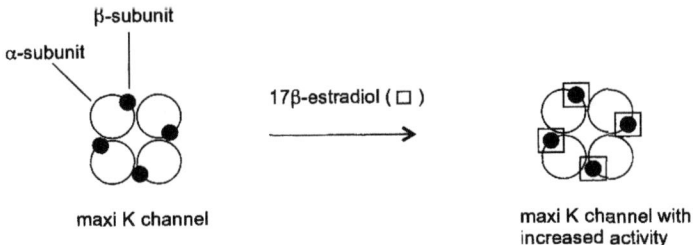

Figure 6.15 *Binding of 17β-estradiol to the β-subunit of a maxi K channel*

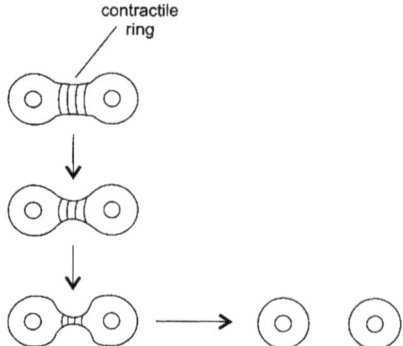

Figure 6.16 *Cytokinesis. Following mitosis (nuclear division), cytokinesis takes place. The contractile ring (actin filaments and myosin II) tightens, pulls the plasma membrane inwards and finally pinches it into two. The ring then disperses. The myosin light chains are regulated by phosphorylation (catalysed by Ca^{2+}-calmodulin associated with myosin light chain kinases) as in smooth muscle*

important steps, both of which are under the influence of Ca^{2+} ions. Actomyosin is much less structured in non-muscle cells.

6.6 ION CHANNELOPATHIES ASSOCIATED WITH MUSCLE DYSFUNCTION

Clearly the malfunction of any of the many ion channels that are used in the complex functioning of all muscles could be responsible for the onset of a disease. We have already alluded to myotonia arising from mutations in Cl^- channels. Myotonia results from a *train* of action potentials, lasting seconds, initiated by a single stimulus rather than from a single action potential (fraction of a second) as with normal muscle. Incomplete muscle relaxation and stiffness are the result. A number of ion channel defects (ion channelopathies) associated with hereditary diseases of skeletal and cardiac muscles in humans and other animals have been described. These result from mutations in ion channels. Clinical studies of these diseases have helped in understanding channel functions and behaviour more fully. Conversely, understanding the molecular basis for the dysfunction could lead to improved diagnosis, treatment and the design of drugs to help combat the disease. In the last decade, more than twenty-five nerve (which have been discussed above) and muscle disorders have been identified as being due to ion channelopathies. Many of the channel disorders in humans cause sudden attacks of illness in otherwise healthy individuals. These diseases often arise from a single amino acid substitution in the mutant form of the channel in different locations (Figure 6.17). The mutations can modify the structure, activity and function of the channel. Selected muscle diseases caused by mutations in ion channel genes are presented in Tables 6.3–6.6.

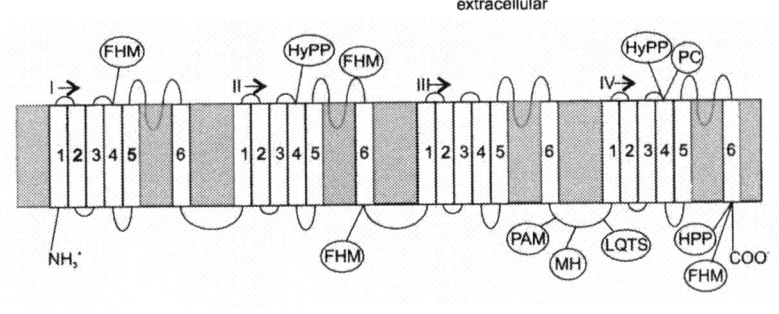

Figure 6.17 *Locations of mutations in the α-subunit of Na and Ca channels in humans leading to diseases. For Na channels, the most frequently encountered resulting in PAM, PC and HPP (human skeletal muscle) and LQTS (cardiac muscle) are shown. For Ca channels, mutations leading to hemiplegic migrane, FHM (neuronal) (Table 4.2) and HypoPP and MH (skeletal muscle) are shown. Malignant hyperthermia also arises from mutations in RyR1. Most known mutations are single amino acid changes*

Table 6.3 *Mutations in voltage gated Na channels in skeletal muscles*

Channel Gene	Result	Disorder(s)
SCN4A Encoding α-subunit of human muscle Na channel. First disease-causing ion channel mutation detected (1991).	Abnormal late opening of Na channels. Prolonged muscle fibre membrane depolarization and delayed relaxation after muscle contraction.	Many are associated with skeletal muscle myotonias (hyperexcitability or weakness).

Table 6.4 *Mutations in RyR in the SR of skeletal muscles, Ca²⁺ ion release channels*

Channel Gene	Result	Disorder(s)
RyR1 Defects are point mutations *e.g.*, R615C (pig) and R614C (human). >20 disease-causing mutations have been identified in humans.	Alteration of receptor structure that prevents proper closing of channel. Prolonged release of Ca^{2+} ions from SR and sustained contraction.	*Malignant hyperthermia* (MH) Genetic predisposition to react abnormally to certain drugs.

Table 6.5 *Mutations in Ca²⁺ ion channels in skeletal muscles*

Channel Gene	Result	Disorder(s)
CACNL1A3 Encodes α-subunit of voltage gated L-type Ca channel (three point mutations). R1086H also reported associated with malignant hyperthermia (above).	Still uncertain; possibly rapid inactivation of channel.	*Hypokalemic periodic paralysis* (HyPP) Episodes of muscle weakness in teenagers. Associated with low serum [K⁺]. Also occurs in Burmese cats!

Table 6.6 *Mutations in the major Na⁺ ion channel in cardiac muscle*

Channel Gene	Result	Disorder(s)
SCN5A α-subunit. Several mutations have been observed. Most severe is a three amino acid deletion.	Channels exhibit an increased tendency to enter second gating mode and long, late openings.	*Long QT syndrome 3* Also idiopathic ventricular fibrillation. Abrupt loss of consciousness, seizures and death.

6.6.1 Myotonias

There are a variety of myotonias associated with point mutations in the α-subunit of a voltage gated Na channel in skeletal muscle (Figure 6.17). These are not life-threatening but are obviously very distressing. They include:

- *Potassium activated myotonia (PAM).* This is brought on in susceptible individuals by the ingestion of potassium rich foods (*e.g.* bananas) and ranges from mild to severe. Replacement of a single glycine residue (by valine, G1306V) in the inactivation gate, III S6–IVS1 linker (Figure 6.17), produces a mild form of the disease whereas replacement of the same residue by a charged glutamate residue with a long side chain (G1306E), leads to severe permanent myotonia. The treatment for the severe form of PAM is a Na channel blocker (*e.g.* tocainide (20)) that moderates the effect by inhibiting abnormal late openings of mutant channels and thus prevents repetitive firings.

(20)

- *Hyperkalemic periodic paralysis (HPP)*. Potassium rich foods and physical exertion can trigger this condition.

Some Quarter Horses May Fall Over After Eating Alfalfa!

Selective breeding of the American thoroughbred quarter horse has meant that as many as 1 in 50 can have an E-HPP mutation where a phe to leu substitution occurs in the S3 domain of repeat IV of the horse skeletal muscle sodium channel. This leads to impaired inactivation and long bursts of opening of the channel, a sustained depolarization and resultant paralysis. The paralysis is induced by feed containing a high concentration of K^+ ions. Injection of a solution of K^+ ions is used clinically to determine if a horse carries the defective gene. It has been suggested that the increase in extracellular K^+ ions occasioned by the ingestion could set off a chain of events leading to sustained cell membrane depolarization, loss of electrical excitability and thus muscle paralysis.

- *Paramyotonia congenita (PC)*. This condition is cold induced, the reasons for which are unclear. Unfortunately eating an ice-cream can temporarily cause slurred speech in an afflicted individual.

Much more serious are defects in calcium channels in skeletal muscle (malignant hyperthermia) and in a variety of sodium and potassium channels in cardiac muscle, *e.g.* in LQTS.

- *Malignant hyperthermia (MH)*. This is an abnormal response to muscle relaxants and general anaesthesia that may be present in approximately 1 in 20 000 adults. When a general anaesthetic is administered to a genetically predisposed patient, the pulse rate increases, skeletal muscles stiffen and the body temperature rises (sometimes to 109 °F!). Without treatment death can occur within minutes. In 1960 the condition was recognized in an Australian family in which ten members had already died after receiving a general anaesthetic. A similar disorder occurs in domestic pigs when stress (*e.g.* approaching an abattoir!) can induce sudden death, which causes undesirable characteristics in the meat. The genetic basis of malignant hyperthermia (MH) was first discovered in pigs. All mutations that cause MH are located in the cytoplasmic domain of RyR1. Although less common, MH can also arise from a mutation (R1086H) in the cytosol loop linking domains III and IV in skeletal muscle calcium channels (Figure 6.17). These mutations lead to increased and sustained $[Ca^{2+}]_i$ under anaesthesia. This causes contractions, increased metabolism, depletion of ATP and the overproduction of heat. Fatalities in humans have been significantly reduced by vigilance for the condition and the use (in humans) of dantrolene (21), which inhibits the release of Ca^{2+} ions from the SR. The drug is now kept in operating theatres for use in emergencies. Disease causing mutations in the related IP_3 receptor, another intracellular ligand gated calcium channel (Table 3.2), have so far only been identified in mice.

(21)

• *Long QT syndromes (LQTS)*. An electrocardiogram is the sum of the many electrical activities in various cells of the heart. The QT interval observed in the electrocardiogram is the period in a heartbeat when the heart muscle is recovering from one contraction and before it is triggered to contract again. Persons with long QT syndrome (LQTS), around 1 in 10 000, have an approximately 0.1 second lengthening (to 0.48 s) from normal, although this is not an accurate method of diagnosis. This usually causes no problem, but the syndrome predisposes the person to cardiac arrest. In fact, for one-third of patients who die from LQTS, death is the first and only symptom of the condition. A number of distinct genetic mutations that encode for voltage gated sodium or potassium channels in the cardiac muscle give rise to congenital forms of LQTS. One of the most severe forms (LQTS3) is the ΔKPQ mutation that occurs in a highly conserved portion of the III–IV inactivation linker in the α-subunit of a sodium channel (Figure 6.17). It is caused by the sodium channel SCN 5A gene and results in the deletion of three amino acids (lys-1505, pro-1506, gln-1507). This portion is normally responsible for the fast inactivation and the mutation causes the Na channel to stay open even after depolarization. This delays the action potential and the QT interval is lengthened. The drug mexiletine (22) is used to block these 'leaky' sodium channels and shorten the QT interval in LQTS3 patients. Some mutations also reduce the number of active potassium channels with the same undesirable result. LQTS can also be triggered by non-genetic means including excess alcohol consumption and certain drugs.

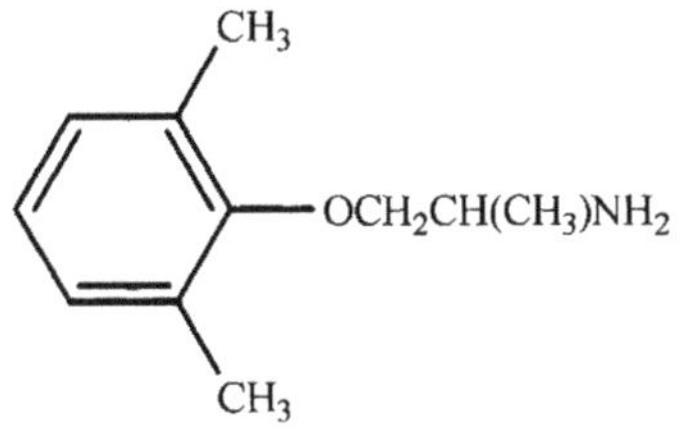

(22)

6.7 SUMMARY

A vital and omnipresent organ is the muscle. It was chosen to illustrate many of the tasks that the s-group ions perform in animals. Muscle consists of an array of fibres that can contract and in doing so produce movement of many parts of the body. Three main types of muscle are used to do this; skeletal, cardiac and smooth. All muscle contractions take place by the sliding of thick and thin filaments past each other, although the particulars differ among the three types. The final events leading to all muscle contractions (a continuation of signal transfer described in Chapters 4 and 5) were detailed. Again different ions (but Ca^{2+} ions dominate) and various channels are involved. Diseases of skeletal and cardiac muscles arising from mutations in ion channel genes can occur. For different reasons these can lead to sudden death in humans, fainting goats, falling quarter horses or stressed-out pigs!

CHAPTER 7

Senses

7.1 TYPES OF SENSATIONS AND THEIR GENERAL PROCESSING

'Nothing is in the mind that does not pass through the senses' (or words to that effect) – Aristotle.

The traditional five senses are touch, hearing, sight, smell and taste. Other senses include pressure (related to touch), temperature and pain. These sensations, from both external and internal sources, will be detected and processed by an organism, which will then take appropriate action. The senses thus enable an organism to react to the outside world as well as to monitor its internal well being. The stimuli can originate in the skin, muscle, bone or associated organs (*somatic sensation*) or be unconsciously controlled in the internal thoracic or abdominal cavities (*visceral sensation*). The general sequence of events from beginning to end is shown in Figure 7.1 and the types of senses are given in Table 7.1.

The sensory cell both detects and then transduces the stimulus, that is it converts the non-electrical signal (*e.g.* light or touch) to an electrical one. In all cases the stimulus, directly or indirectly, alters the activity of an ion channel(s) in specialized receptor membranes and generates (sometimes after several steps) an action potential. The ion channels are often

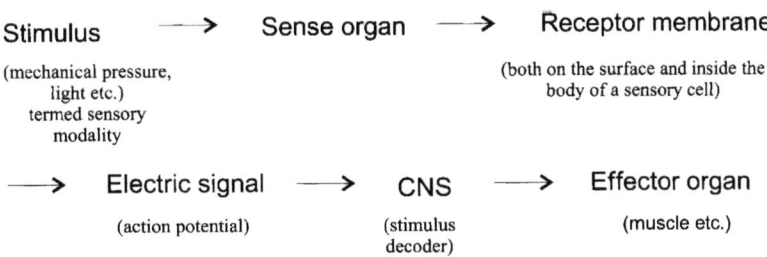

Figure 7.1 *The stages linking a stimulus to a response*

Table 7.1 *Sensory systems*

Sensation	Sense Organ	Receptor Cell
		Mechanoreceptor
Pressure	skin and deep tissue	encapsulated nerve endings
Touch	skin	nerve terminals
sound (hearing)	inner ear (cochlea)	hair cells
balance	vestibular organ	hair cells
muscle stretch	muscle spindle	nerve terminals
		Thermoreceptor
cold, heat	skin, hypothalamus	nerve terminals and central neurons
infrared heat	facial pits of rattlesnake	nerve terminal
		Photoreceptor
light (vision)	eye (retina) and pineal gland (birds)	photoreceptors
		Chemoreceptor
Taste	tongue	taste bud cells
Smell	nose	olfactory receptors
		Nociceptor
pain (mechanical, thermal or chemical)	skin and various organs	nerve terminals

nonselective or at most, weakly selective. Messages from the sensor arrive at different places in the CNS where they are processed differently, depending on the nature of the sensation. Two or more sensory inputs can occasionally use the same receptor (*e.g.* capsaicin and heat). There appears to be little doubt that the senses are continuous. For example, a combination of the odour of lemon with a yellow colour produces more brain activity (measured by functional magnetic resonance imaging) than does presentation of the odour and the colour red to a subject. Table 7.1 shows the types of sensations that are likely to be encountered by most animals. The receptors for visceral sensations are the same as those for somatic sensations, but each sensation is associated with a specific receptor and does not perturb other receptors. Simple diagrams of some sensory receptor cells are shown in Figure 7.2.

Sensory receptors respond to both external (exteroceptor) and internal (interoceptor) stimuli. The stimuli are converted, *via* ion channels, to a receptor potential that is usually a depolarization (but can be a hyperpolarization) of the membrane. Some sensory cells (*e.g.* mechanosensors) are neurons and if the depolarizing receptor potential exceeds threshold, an action potential will result. An example is the crayfish abdominal muscle stretch receptor.

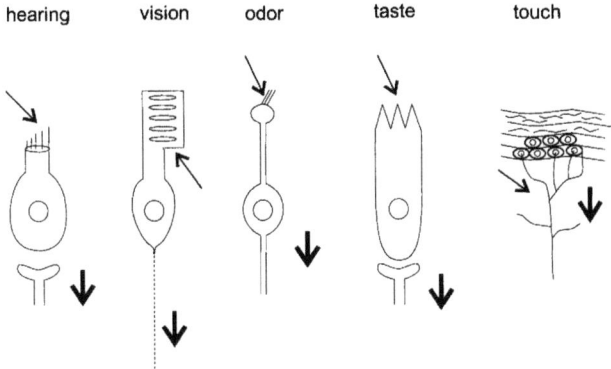

Figure 7.2 *Sensory receptor cells in vertebrates. The light arrows indicate the sites on which stimuli act and the heavy arrows indicate the sites of generation of action potentials. The touch receptor is only one of several receptors in the skin*

Crayfish Have Mighty Tails!

The tail muscles of crayfish have stretch receptors that consist of sensory neurons. The neurons have dendrites that are stretch sensitive and are embedded in a special bundle in the tail muscle. When a crayfish tail bends, the muscle is stretched and the receptor is activated. This results in graded potentials in the soma that are converted to action potentials in the spike-initiating zone (near the first node of Ranvier, Figure 4.3). The magnitude of the stimulus is reflected in the strength of the receptor potential and in the frequency of the action potential. At the terminus of this sensory neuron neurotransmitter release invokes a graded potential in the next (second-order) neuron. If this reaches threshold, an action potential is induced and the process is repeated.

Other sensory cells (*e.g.* vertebrate photoreceptor) lack axons, but contain vesicles that release transmitters onto afferent neurons (Figure 4.1). One or more synaptic relays are essential before the required action potential can be generated. Receptor potentials on short receptors promote the release of neurotransmitters onto adjacent neurons (*e.g.* taste and vision). Receptor potentials on long receptors cause depolarization and action potential discharge *via* second messenger systems that activate cation channels (*e.g.* odour, touch and pressure concentrated in fingertips, as well as stretch receptors in vertebrate and invertebrate muscles).

7.2 TOUCH AND HEARING

There are a diversity of mechanosensors distributed throughout the mammalian body that respond to touch, sound and pressure. If a thin plastic rod is pressed against a finger, action potentials are recorded at a single pressure-sensitive mechanoreceptor located on the finger. Increasing the force on the rod (increased stimulus strength) registers as an increased receptor potential which in turn increases the action potential frequency, but not its magnitude, in the afferent neuron. Mechanoreceptors are abundant in the mouth and on fingertips. The fingertips are more sensitive than the back of the hand where there are fewer mechanoreceptors.

The receptor protein incorporates an ion channel that opens immediately on stimulation (stretching of the incorporating membrane) by undergoing a conformational change. Many of these stretch-activated channels are permeable to both univalent and bivalent cations and therefore can participate in the regulation of osmotic pressure changes within cells. The conversion of the small mechanical movement (as little as 0.1 nm in some cases) into the opening of the channel and firing of the axon is not easy to understand.

Sensory cells that respond to physical movement are present in a wide variety of creatures from the most simple (those without a nervous system) to the most complex animals. A few examples have been selected.

7.2.1 Paramecia Respond to Touch

Paramecia are single-celled organisms, approximately 150 μm in length and slipper-like in shape, which do not have neurons nor muscles, but do possess stretch-activated channels. The creatures have hair-like outgrowths on their surfaces called *cilia*. When thousands of cilia point in the posterior direction and move in coordinated waves ('beating') the paramecium moves forward (and *vice versa*). A touch on the anterior end (*e.g.* by collision) causes the organism to reverse direction. The touch opens Ca^{2+} ion selective stretch activated channels that are located in the ciliary membrane. Receptor potentials are generated that depolarize the membrane of each cilium, which of course is negatively charged in the resting state. There follows a Ca^{2+} ion initiated action potential. At sufficiently high $[Ca^{2+}]_i$ the beat of the cilia is reversed and the creature reverses for a short time. At this point, various Ca^{2+} ion activated potassium channels open, K^+ ions leave the cell, which repolarizes, the Ca channels close and the original beat is resumed. Ca^{2+} ATPases may also pump out Ca^{2+} ions. When bumped at the rear end, stretch-activated potassium channels in the posterior membrane open. The cell becomes hyperpolarized and (somehow) the beat intensifies leading to faster forward motion and escape! The intraciliary calcium concentration controls the velocity and direction of swimming. Forward speeds are fastest at $[Ca^{2+}]$ ~10 nM. The creature stops swimming at a $[Ca^{2+}]$ of around 1 μM and reversal maximum is reached around 50 μM. Some mutants, *e.g.* the pawn, (the chess piece that cannot move backward) have no Ca^{2+} ion channels and cannot take evasive action. Other mutants have fewer potassium channels and reverse excessively.

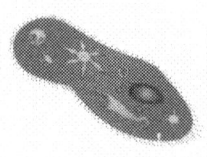

Paramecia Could be Used to Evaluate Antimalarial Drugs

Antimalarial drugs such as quinine (7.5.3) inhibit the calcium-dependent backwards swimming paramecia noted above, but are toxic to the creature at high concentrations. Pawn mutants or deciliated paramecia that lack calcium channels are much less susceptible to antimalarials than are the wild type. Calcium channels could therefore be the sites of the toxic action of the drugs not only in paramecia, but also in other protoctista such as the parasites that are responsible for malaria and amoebic dysentery. Such small creatures can therefore be important in drug research.

7.2.2 Cockroaches Respond to Wind

Many of the signalling mechanisms that are used by the cockroach to move away from a potential attacker in a specific direction are well understood. A simple representation is shown in Figure 7.3. Many hairs in the posterior cercus of the cockroach are deflected by nearby wind movement caused by, for example, a predator toad's tongue! This deflection excites the sensory neuron at the base of the hair. These wind receptor neurons are in contact (synapse) with a giant interneuron (one of fourteen) in the terminal ganglion of the nervous system. This then synapses, in the thoracic ganglion, with one of a few motor neurons that control the leg muscles. Wind movement near the posterior will therefore promote rapid movement of the leg muscles and thus effect escape. Cockroaches whose cerci have been covered with wax have a nearly tenfold increased chance of being caught.

7.2.3 Spiders Respond to Vibrations

A spider is alerted to the presence of prey trapped in its silken web by vibrations acting on a mechanoreceptor organ in the spider's legs. An even

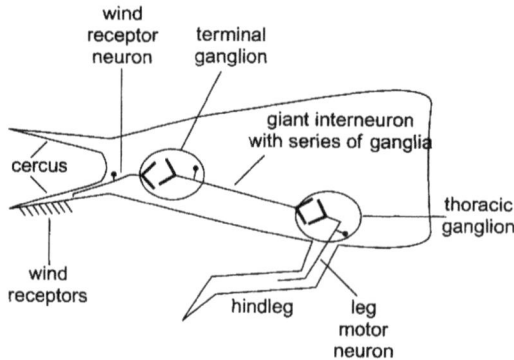

Figure 7.3 *Escape mechanism of cockroach Periplaneta americana using wind receptors*

more elaborate example of this effect involves the nocturnal desert scorpion *Paruroctonus mesaensis*. Vibrations in the sand caused by prey transmit to a particular set of sensitive mechanoreceptors on the eight legs of the scorpion. Action potentials arising in the different legs are analysed by the CNS that allow the creature to estimate the direction and location of the disturbing presence over a distance of up to 15 cm.

7.2.4 Alligators Respond to Ripples on the Surface of the Water

Small dark dots on the faces of alligators are pressure receptors that are innervated by a thick cranial trigeminal nerve. When young alligators were half submerged in a tank of shallow water, a slight disturbance of the water surface evoked a robust reaction from the crocodilian. A spike firing of the nerves attached to the facial dots was observed, which increased in frequency with increased wave amplitude. It is believed that these sensory organs are used to detect even small disturbances on the water surface to aid in nocturnal hunting.

7.2.5 Humans Respond to a Tap on the Knee

Anyone who has visited the doctor's surgery will have encountered the knee jerk test! This reflex beautifully illustrates the route from stimulus to result (Figure 7.4). The sense organ of the receptor (muscle spindle) and the effector organ (muscle fibres) are associated with the same muscle. A

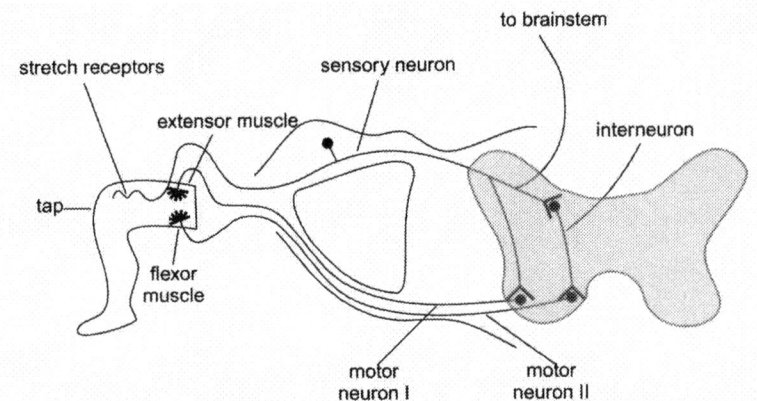

Figure 7.4 *Knee jerk components. A sharp tap just below the kneecap causes a tendon reflex. The normal response is a quick upward jerk of the knee and indicates that all components (neurons, neurotransmitter junctions etc.) are functioning correctly. An absent or a violent reflex can indicate a disease of the nervous system*

tap on the knee tendon just below the kneecap stretches the thigh muscle to which it is attached. Stretch receptors (muscle spindles) incorporated in the thigh muscles are stretched and activated producing a receptor potential in the sensory terminal. An action potential is fired in the sensory neuron. This neuron synapses with the motor neuron I and, *via* an interneuron, with motor neuron II. It is here that the synaptic potentials are generated. Motor neuron I transmits an action potential to extensor muscles that contract, thus shortening the thigh muscles and producing the knee jerk within 50 ms. However, this is not the whole story. Activation of the interneuron *inhibits* motor neuron II, which controls the flexor muscle. The *relaxation* of this muscle is necessary because extensor muscles and flexor muscles work in antagonistic pairs, one contracting as the other relaxes. The response is involuntary, but some contacts with the brain must be present, since the 'tapped' person is aware of the jerk. The route through neuron I is *monosynaptic* and is termed a *stretch reflex*. *Polysynaptic* transmissions involve neuron II.

7.3 HEARING IN HUMANS

We humans can distinguish hundreds of thousands of different sounds. A special mechanoreceptor in the hearing apparatus of the human ear responds to sound and balance. Deep inside the inner ear, almost behind the eye, lies the *cochlea*. This is the snail shaped organ containing sensory hair cells that are mechanoreceptor ion channels. These are the sensory cells of the human auditory system (Figure 7.5).

hair bundle

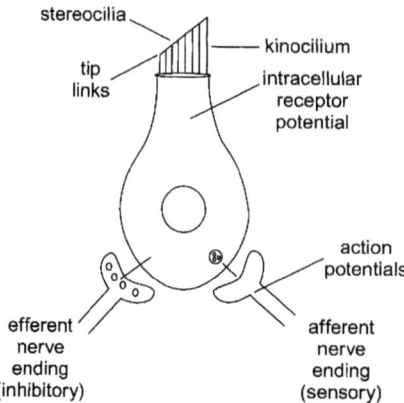

Figure 7.5 *An inner hair cell sensory receptor. It responds to waves of pressure (sound) and transmits the information using afferent nerves. There are also (more numerous) outer hair cells that are the postsynaptic targets of cholinergic efferent neurons that project from the brain stem. Efferent feedback has a variety of functions including protection from damage from loud noises (increasingly valuable in the modern world)*

   Stereocilia are projections on the hair cells and near their tips (*tip links*) are the mechanoreceptors that respond to vibrations arising from sound waves collected and transmitted from the outer and middle ear. The fluid inside the cochlea (endolymph) is used as a transmitting medium. The waves in the fluid cause vibrations in the membranes on which the sensory hair cells sit. At rest, about 15% of the channels in the hair cell are open and the resting potential is approximately −60 mV. When the hair cell bundle is deflected toward the kinocilium (by even 1 nm!) the nonselective ion channels in the plasma membrane of the stereocilia open very quickly (20–200 μs). This allows K^+ ions to move from the endolymph ($[K^+]$ ~150 mM, $[Na^+]$ ~1 mM) into the hair cells. These relative concentrations of K^+ and Na^+ ions are unusual for an extracellular fluid (see Table 1.1). As a result of the K^+ ion influx the membrane is depolarized by some 10–20 mV (this depolarization is equivalent to the effect of Na^+ ions in neurons). The depolarization opens calcium channels, which promotes the release of neurotransmitter (probably glutamate) on to the dendritic ends of the sensory neurons. Thus an action potential is set up in the axon forming cochlear nerve (not the hair cell itself). This carries the signal to the CNS where it is recognized as sound. Each hair cell and associated nerve fibre responds best to a single frequency of sound. Displacement of the bundle away from the kinocilium closes an ion channel. This causes hyperpolarization and a decrease in transmitter release.

Rapid recycling of K^+ ions into the endolymph is essential for normal hearing. Six of the many proteins involved in K^+ ion recycling, when malfunctioning, are associated with certain kinds of deafness in humans (and mice). A voltage gated potassium channel encoded by the KCNQ4 gene in subtypes of hair cells and neurons of auditory and vestibular systems is thought to transport K^+ ions out of the cell. Persons with mutant forms of this gene experience progressive hearing loss.

The vestibular organs in the inner ear help maintain balance and are connected to the cochlea. A gelatinous substance in which tiny stones (otoliths, particles of inorganic material, 8.4.1) are embedded covers the stereocilia in the vestibular system. Motion causes the gelatinous–otolithic material to change position, resulting in a pull against the hair cells and stimulation of the receptor cells. This is used to give information about head movement and its position relative to gravity.

Fish Use Hair Cells for Synchronized Swimming

Fish and amphibians have, on either side of their bodies, a set of external sensory cells called the *lateral line system*. The system is based on hair cells and is sensitive, *via* a covering called the capula, to motion in the surrounding water. It enables the fish to respond to the rapid movement of predators or of prey and to direction changes of neighbouring fish, to thus maintain the beautiful synchronized swimming seen in schools of fish. Hair cells in the lateral lines of some fish, *e.g.* catfish, have lost their cilia and have been modified to become electrosensors to detect electric currents (7.4.3).

7.4 SIGHT

Photoreceptors are sensory cells that respond to light. They are widespread in the animal kingdom. They are found even in unicellular organisms that react to light, but the processing system increases in complexity until the sophisticated image-forming vertebrate eye is reached. The conversion of a visual to an electrical signal is probably the best understood of all sensory mechanisms. The eyes of vertebrates differ only in detail and the human eye is a good model.

7.4.1 Vertebrate Eye

Light enters a vertebrate's eye through the cornea, passes through aqueous humor, pupil, vitreous humor and then the front surface of the retina. It finally reaches the photoreceptor cells called rods (Figure 7.6A) and cones. These process low intensity light and high intensity light (and colour), respectively. The photoceptors contain photopigments that absorb light. Rods contain only one photopigment, rhodopsin, consisting of opsin protein and bound chromophore 11-*cis*-retinal (1), a derivative of vitamin A. Cones have three different opsins, but still only the one chromophore, retinal. Because rods are more readily available, they have been the more studied.

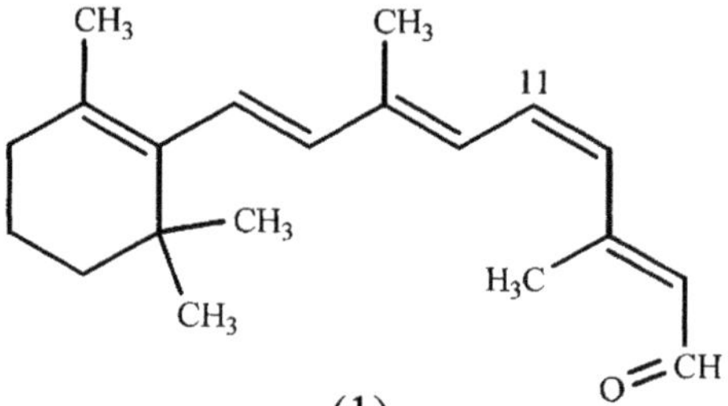

(1)

7.4.2 Conversion of Light to Electrical Signals

Briefly, nerve signals as slow membrane potential changes arising in the rod and cone cells, pass to bipolar nerve cells. They then proceed to ganglion cells where a sequence of action potentials is initiated (Figure 7.6B). In order to accomplish this, G-proteins are used to mediate signal transduction (Figure 7.7 and Table 7.2).

Rhodopsin is a membrane-spanning protein containing seven α-helices. It is a conjugate of opsin, retinal and a phospholipid. Retinal sits in the centre of opsin. Rhodopsin is a G-protein coupled receptor. It is a member of a large family of membrane proteins that respond to many types of stimuli (Table 7.2). Rhodopsin is the first such protein whose X-ray structure has been solved (in 2000). The specific G-protein used in light sensing is called transducin. Light converts 11-*cis*-retinal (1) to all *trans*-retinal (by rearranging around the double bond at position 11; several intermediates are involved). Within 10 ms this conversion induces a conformational change in rhodopsin and converts it from the inactive to the active form. Active rhodopsin activates transducin, the α-subunit of which stimulates the activity of a cGMP phosphodiesterase. This leads to

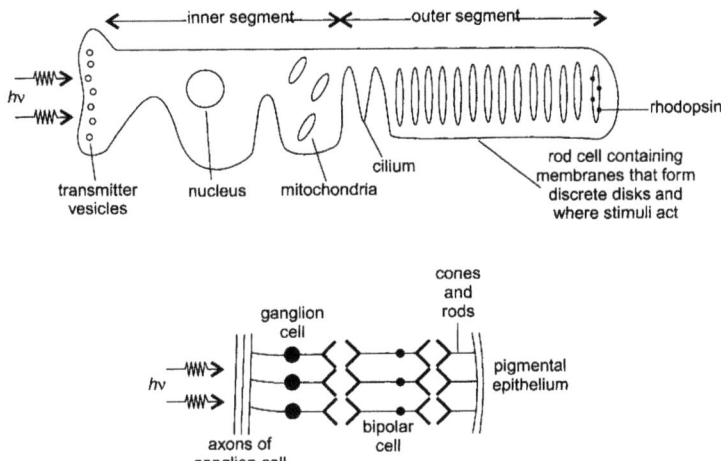

Figure 7.6 *(A) Rod-shaped sensory cell in vertebrate retina (top). (B) Transduction of light stimuli into neuron signals in vertebrate retina. A number of other nerve cells add to the network shown and aid in the preprocessing and codification of information before it reaches the optic nerve*

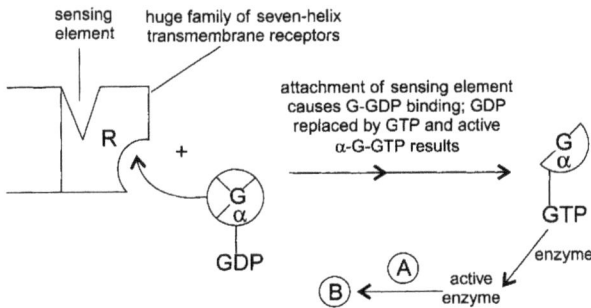

Figure 7.7 *G-protein mediated transduction. Binding of the sensing element induces a change in the shape of R that activates the G-protein. The conversion of A into B can effect the gating of a cyclic nucleotide gated ion channel or lead to neurotransmitter release (see Table 7.2)*

a *decrease* in intracellular cGMP concentration as it is converted to 5'-GMP within about 100 ms of the light flash. The cGMP bound to an Na channel is directly responsible for holding the channel open and therefore the maintenance of a depolarized cell. The loss of cGMP leads to closing of the Na channels and *hyperpolarization* of the cell. There is a consequent reduction in the rate of glutamate (transmitter) release at the synapses of

Table 7.2 *G-Protein mediated transductions with a variety of sensors*

Sensing Element	Receptor (R)	G-protein	Enzyme	A	B	Result
Light (*hv*) acting as 'neurotransmitter'	Rhodopsin	Transducin	Phosphodiesterase	cGMP	5'-GMP	Cyclic nucleotidegated (CNG) ion channel closes.
Odorant molecule	Olfactory	G_{olf}	Adenylate cyclase	ATP	cAMP	CNG ion channel opens and action potential results.
Sweet taste	Taste	Gustducin	Adenylate cyclase	ATP	cAMP	K channel closed by phosphorylation; depolarization as K^+ ions accumulate in cell.
Bitter taste	Taste	Gustducin	Phospholipase C δ1	PIP_2	IP_3	Ca^{2+} released from stores; neurotransmitter release.

the rod cell connection to the bipolar cells. This decreased release of glutamate causes the bipolar cells to experience a change in membrane potential that registers as an altered signal to the brain. Although action potentials are *not* fired in photoreceptor cells, they are finally produced in the ganglion axons of the neurons making up the optic nerve. When the light stimulation ends, cGMP is regenerated and the receptor returns to its original state.

7.4.3 Electroreceptor Cells

Only a few fish (electric eel, electric catfish) can produce lethal electric shocks. However, many fish (*e.g.* the knife fish) in the fresh waters of Central America use complex weak electrical discharges to locate enemies or prey and to navigate and communicate with each other in murky water since they have poorly developed vision. Each individual in a group emits electric pulses at slightly different frequencies and adjusts this when a new fish arrives! Such weakly electric fish use a specialized electric organ in their tails to generate electric pulses that are then sensed by electroreceptor cells distributed throughout the head and body of the animal. These cells make synaptic contact with certain nerve axons. The result is a train of current pulses flowing through the surrounding water, from the posterior to the anterior end of the fish. Calcium channels, synaptic transmitters and action potentials all take part in this process.

Duck-billed Platypus Attacks Defenceless Battery

During a dive, a duck-billed platypus (*Ornithorhynchus platinus*), in a reflex action, closes its eyes, ears and nose and relies on electroreceptors (in addition to mechanoreceptors) in its bill to locate prey. Experiments have shown that the animal can home in on an electric dipole (2 mV cm^{-1} from a 1.5 V battery at a distance of 10 cm) and seize the source with rapid snaps of the bill. This field strength is similar to that generated by the tail flicks of an escaping freshwater shrimp (a potential meal) providing a guide to the platypus of its whereabouts. It has been demonstrated that action potentials are evoked in the cortex of the platypus during the tail flick of the shrimp.

7.5 SMELL AND TASTE

Chemical compounds activate chemoreceptors. Some chemicals are from external sources, *e.g.* those associated with smell and taste. Other chemicals such as O_2 and H^+ generated internally are sensed by receptors in the carotid body, which thereby coordinates changes in the composition of the blood. The chemoreceptors in marine animals are very sensitive to amino acids, even at concentrations as low as nM. This helps in the location of their food. The spiny lobster *Palinuris*, for example, has chemoreceptors in antennules that can detect, using an electroolfactograph a concentration of only 0.1 nM of taurine ($H_2N(CH_2)_2SO_3H$). Considering the high concentrations of salts in the environment, this is a remarkable achievement. Analyses of tissue extracts from marine molluscs and fish (the main food of lobsters) show that taurine is one of the most prevalent amino acids. It is difficult to decide whether one should speak of a fish smelling or tasting its food. Smell and taste are closely related. Humans have difficulty in tasting food when suffering from a blocked nose.

7.5.1 Vertebrate Olfactory Receptor

This receptor is also a neurosensory cell because it conducts action potentials. There are about 1000 different receptor proteins, each encoded by its own gene, with specific sensitivity to odorants. Even enantiomeric forms can be distinguished. To most people (+)-carvone has a caraway smell whereas (−)-carvone has a minty odour. This indicates that human olfactory receptor sites are chiral. Humans can recognize approximately 10 000 scents.

Olfactory receptor proteins have seven hydrophobic membrane spanning domains and are members of the G-protein-coupled receptor family. The greatest diversity is in the 3–5 spanning domains and these could contain the olfactory binding site(s). Receptors that are used for vision (7.4.2) and for sensing sweet and bitter tastes (7.5.3) also have seven transmembrane α-helical domains and require G-protein transduction (Table 7.2).

Odorant molecules are either transported by proteins or diffuse through the mucus lining and are delivered to the cilia (Figure 7.8). In either case, they bind to olfactory receptor proteins in the ciliary membrane of the cell. A cAMP cascade is initiated by this binding *via* a G-protein adenylate cyclase and ATP. This cascade generates (Table 7.2) action potentials *via* opened cyclic nucleotide gated (CNG) ion channels. A Na^+ and Ca^{2+} ion influx results in depolarization and an action potential. In addition, the Ca^{2+} ions can activate a Ca^{2+} ion-dependent Cl^- channel and because (unusually) intracellular $[Cl^-]$ is high there is an efflux of chloride ions,

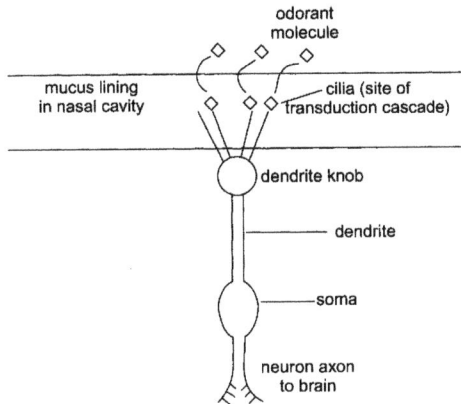

Figure 7.8 *Olfactory receptor cell. A vertebrate olfactory receptor cell, embedded in a supporting cell, is a bipolar neuron with a single dendrite at the tip of which are a number of cilia. A single axon projects to the olfactory bulb of the forebrain right behind the nose. Insect and crustacean olfactory receptors, which have been most studied, are similar*

which enhances depolarization. The axons of the olfactory unmyelinated neurons (there are hundreds) synapse with relay cells in the olfactory bulb in the CNS and thence produce messages in the brain. Other transduction pathways almost certainly exist involving, for example, Ca^{2+} and/or IP_3 activated cation channels in some vertebrates and invertebrates. The olfactory epithelium in the nasal cavity where odour is sensed is only a few square centimetres in area in humans, but is more than $100\ cm^2$ in dogs. Dogs have been shown to be able to detect cedarwood odour at a concentration of only 10^{-17} M. Whales and porpoises have no olfactory sensors and consequently cannot smell. Mice whose gene encoding for the α-subunit of the CNG channel has been 'knocked out' quickly die because they cannot smell (and suckle). Zebrafish are proving to be useful to neuroscientists. They have a simpler olfactory system and basic neuronal circuitry than do mammals. In addition, they are the first vertebrates to prove amenable to large-scale genetic screening as has been done with fruit flies and worms.

Salmon and Trout Tolerate Both Fresh and Sea Water

Salmonid fish hatch in fresh water, migrate to the oceans and return to their birth river to reproduce. Their olfactory epithelium, which is exposed directly to the external environment, has therefore to adjust to widely different ionic conditions. In seawater the concentrations of Na^+ (500 mM), K^+ (10 mM) and Ca^{2+} (10 mM) ions are high enough to allow the generation of a depolarization leading to the desired action potential. In contrast, in fresh water these ion concentrations are low (barely mM) and can hardly promote depolarization by cation influx. However L-amino acids (in concentrations as low as nM) can trigger Ca^{2+} ion influx. An increased $[Ca^{2+}]_i$ will activate a Ca^{2+} ion activated Cl^- channel and a Cl^- ion efflux can generate the necessary depolarization in the olfactory receptor neuron. These fish rely on outside Ca^{2+} ion concentration to allow the olfactory receptors to recognize amino acid stimuli (see above). Only Ca^{2+} ion (not Na^+ or K^+) is necessary for receptor response in a variety of fish.

7.5.2 Vertebrate Taste Receptors

Vertebrate taste centres, unlike olfactory sensors, are not neurons, but have neuron-like character. It is now generally accepted that taste cells are electrically excitable and therefore carry a negative internal charge. Sensing of stimuli occurs at receptors embedded in microvilla at the surface of taste buds. A taste bud contains 50–100 cells that include precursor, support and taste receptor cells (Figure 7.9).

Binding of tastant (the effective chemical in the food) at the receptor promotes an increase of positive ions inside the taste cell (depolarization) by mechanisms dependent on the nature of the taste modality (see below). Depolarization results in the release of Ca^{2+} ions in the cell through

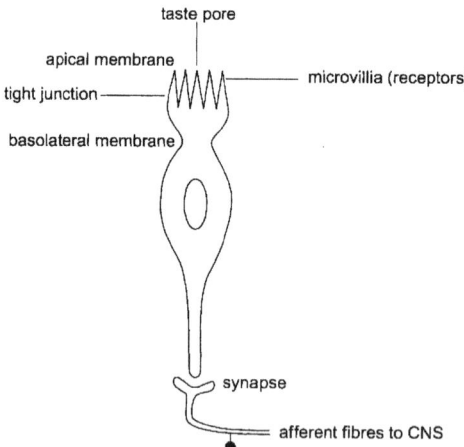

Figure 7.9 *One of many taste receptor cells within a mammalian taste bud. Taste cell receptors are on the apical membrane, which is separated from the basolateral membrane by tight junctions (2.1.2). The basolateral membrane synapses with primary afferent nerve fibres*

voltage-sensitive calcium channels or from the ER where they release neurotransmitter onto sensory nerve terminals. The nerve impulse generated transmits information to the taste centres of the cortex through synapses in the brain stem and thalamus. Messages then proceed to the mouth to accept or reject the chemical. The synaptic events are not well understood and the identity of the neurotransmitters is uncertain. More than one could operate within a taste bud.

In contrast, invertebrate taste receptors (studied mainly in insects) are bipolar sensory neurons. In this they differ radically from those of vertebrates, although many aspects of stimulus transduction are similar.

7.5.3 Taste Modalities

As well as providing enjoyment in eating for mammals, taste can act as a warning signal, *e.g.* bitterness could suggest toxicity (*e.g.* strychnine), while a sour taste can indicate spoiled food. Taste is the least well understood of the human senses.

There are essentially only four taste modalities in different overlapping regions of the epithelium of the tongue. A fifth taste (umami = yummi!), common in Asian cuisines, ascribed to monosodium glutamate (used as a flavour enhancer) is also sometimes included. Signalling pathways and transductions are different for different tastes, but *all lead to depolarization* arising from blocked K^+ ion exit from or enhanced Na^+ ion entry into the cell. The tastes and their favoured modes of action are shown in Table 7.3. Sour and salty tastes result from the influx of ions through channels in the uppermost membranes of taste receptor cells. The other tastes are believed to be detected by binding of the taste molecule to the receptor that is linked to a G-protein (therefore G-protein-coupled receptors). Behaviour could be species-specific. For example in Amphibia, bitter chemicals like quinine block potassium channels and cause depolarization directly and bypass the G-protein route. It should be emphasized that classes of tastes do not rely on a single transduction mode.

A Salty Taste Receptor is Modified by its Environment

As might be expected the taste receptor function is modulated by environmental factors. In sodium-restricted rats the nerve response to NaCl is reduced and this permits foods and solutions that are high in sodium to be ingested. In contrast, nerve response to NaCl (but not to other tastes) in

Table 7.3 *Taste modalities, chemical example and possible modes of action*

Taste Modality	Possible Mode of Action
Sour; H^+ ions	Direct, from influx of H^+ ions that block K^+ ion channels. The increased positive charge depolarizes the cell.
Salty; NaCl	Direct, from influx of Na^+ ions through amiloride blocked and other Na channels in microvilla.
Sweet; sucrose	Specific membrane receptor. Binding promotes G-protein mediated attack of cAMP and phosphorylation of potassium channels, closing them. Artificial sweeteners may use the IP_3 second messenger system.
Bitter; many drugs	Specific membrane receptor. Binding promotes internal Ca^{2+} ion release from the calcium store or through a channel (analogous to visual transduction).
Umami; monosodium glutamate	Specific membrane L-amino acid receptor. Coupled to G-protein and release of $[Ca^{2+}]_i$. The detailed mechanism is unknown.

rats is enhanced when the temperature is lowered from 22 to 4 °C. That a sodium channel is responsible for the effect is shown by its suppression by amiloride, the epithelial sodium channel blocker (3.6.1). The enhancement at low temperatures could be a mechanism by which the animal avoids excess NaCl intake, because its blood pressure is increased at cold temperatures.

Not all Bitters Taste the Same

When the receptor for a bitter taste (recently identified) is activated by a bitter compound, a string of reactions involving a G-protein occurs. This leads to the production of a $[Ca^{2+}]$ spike inside the cell and to neurotransmitter release.

The triggering of an increase in $[Ca^{2+}]_i$ has been used cleverly to shed light on cell response to different bitter compounds. If a taste cell can distinguish among different bitter compounds, only some 'bitters' will promote an internal Ca^{2+} ion boost.

Five different common bitter compounds, such as quinine (2) and cycloheximide (3), were added one at a time to bitter sensitive taste cells from a rat's tongue and the calcium ion concentration was determined using a fluorescent indicator (calcium green 1-dextran). Sixty five percent of the cells fluoresced in response to a single bitter

compound. Approximately 25% of the cells responded to two compounds and only 7% reacted to three or more 'bitters'. It appears that
different taste cells are tuned to different bitter compounds and that
individual taste cells can discriminate among bitter stimuli.

(2) (3)

*(2)(3) Radically different chemicals provide a bitter taste. Two of the compounds used in
the study are shown*

Taste receptors relay signals to the brain through temperature sensitive
nerve fibres. Thermal and gustatory receptors are proximate in the mouth.
For some people, cooling the front of the tongue (from 35 °C to 15 °C)
produces a salty taste while warming it from 20 °C to 35 °C produces a
weak sweet taste. Understanding the mechanisms of taste (and of odour)
transductions will help in the treatment of taste and odour disorders like
ageusia, a complete lack of taste and dysosmia, a spurious odour
sensation.

7.6 HOT AND COLD

The sensing of temperature by warm blooded animals (mammals and
birds) is required, for example, to avoid damaging heat and to regulate
their internal temperatures. Only recently has there been some understanding of the mechanism by which temperature is sensed (particularly of
noxious thermal stimuli – see Pain, nociceptors 7.7) by sensory neurons.
There are two types of thermoreceptors present in the free nerve endings
of somatosensory neurons that inervate the skin and mouths of primates
(humans and monkeys) and subprimates (dogs and rats). *Warmth* receptors
increase their discharge rate from zero on being warmed from 35 °C
(approximately body temperature) to 42 °C and *cold* receptors do likewise
when cooled from about 35 °C to 28 °C. In doing so they signal the
sensation of hot or cold to the CNS.

A striking example of heat sensing is displayed by the pit viper.

Pit Vipers Detect Heat Radiating from Prey

The two facial pits of these snakes contain extremely sensitive infrared receptors within membranes at their bottoms. The pits lie between the eye and nostril and are about 5 mm deep. These receptors can detect the radiant heat (infrared radiation emitted by a mammalian body) from warm-blooded potential prey (*e.g.* a human hand only one foot away!), whose body temperature is only 0.1 °C above ambient. The sensory axons of the pit organs increase their firing rate when the pit temperature increases by only 0.002 °C! The facial pits probably also provide stereoscopic perception and help in the location of prey. Besides the pit vipers (a family of snakes that includes rattlesnakes and copperheads) the *Boidae* family (boa constrictors, pythons) also uses thermoreceptors to sense environmental temperatures. How is this phenomenon observed experimentally? A CO_2 laser produces an infrared emission at 10.6 microns, which is close to that emitted by likely prey (*e.g.* birds and mammals). Thus the laser can be used as a stimulus. The resulting electrical activity in the cerebral cortex of a boa can be picked up by electrodes placed there and amplified using an electroencephalograph.

The cellular and molecular basis of hot and cold detection in the mammalian somatosensory system has been clarified by the use of

(4)   (5)

(4)(5) *Menthol (5) is found in peppermint. A mint plant Mentha piperita when placed on the skin or tongue produces a cooling sensation and is used for the alleviation of joint and muscle pain. Capsaicin (4) is the main pungent ingredient in hot chilli peppers. It is a part of the chemical defence that the plant uses to prevent being eaten by mammals, but does not affect seed-dispersing birds (7.7.2). Chilli peppers have been cultivated in South America for over 7000 years*

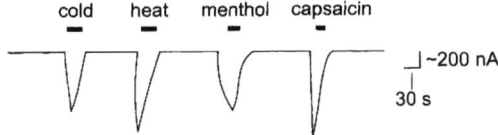

Figure 7.10 *Xenopus laevis oocytes coexpressing CMR1 and VR1 channels. Complementary RNA transcripts were synthesized from CMR1 and VR1 cDNA templates using RNA polymerase. The transcripts were injected into defolliculated oocytes. Two to seven days after injection the electrode clamping recordings shown were carried out. They show that the two channels confer responses to cold (35 °C to 8 °C), menthol (100 μM), heat (25 °C to 50 °C) and capsaicin (1 μM)*
Reprinted with permission from *Nature* from D. Julius, 2002, 416, 52–58, copyright 2002 Macmillan Publishers Ltd.

capsaicin (4) and menthol (5) to simulate the effects of heating and cooling.

A channel vanilloid receptor subtype (VR1) is activated by moderate temperatures above 43 °C as well as by vanilloid compounds, of which capsaicin is a standard bearer. A menthol sensitive receptor (CMR1) on the other hand is activated by cold (8–28 °C, Figure 7.10). Both receptors are members of the same (transient receptor potential, TRP) subfamily of channels. Although they are nonselective toward M^+ ions they have relatively high Ca^{2+} ion permeability. It appears that these TRP channels are the primary detectors and sensors of temperature over a range from about 8–43 °C.

How small changes in temperature can change these channels from a closed to an open state is still unclear. The action is probably direct and requires no subsidiary pathway if the behaviour of nociceptors (7.7) is any guide to heat and cold receptors. Even in cell free membrane patches from nociceptors, action potentials are initiated at around 45 °C.

7.7 PAIN

Nociceptor-specific ion channels are sensitive to injury or pain. Pain is initiated when peripheral terminals of a subgroup of neurons are activated by a diversity of particularly strong sensations. These include mechanical (such as a cut), extreme temperatures or a variety of chemical compounds that are often released as the result of an injury. All are stimuli that have already been considered as minor upsets. The chemical compounds can be external (*e.g.* from a wasp sting) or arise internally (*e.g.* bradykinin or histamine).

7.7.1 Pain Signalling and Pain Moderation

The signalling pathways for pain are different from those exhibited by analogous perturbations such as a gentle touch (Figure 7.11, see also Figure 4.2).

Afferent nerves (thick bundles) synapse to the CNS directly. There are two different afferent axons:

- Myelinated thick (1–4μm diameter) axons, called Aδ, that respond quickly (6–24 m s^{-1}) to the first sign of intense pain.
- Thinner, more abundant (0.1–1.0 μm diameter) unmyelinated axons, called C, which respond more slowly (0.5–2.0 m s^{-1}) to prolonged dull pain induced by chemicals, *e.g.* capsaicin.

In the spinal cord, the ends of nociceptor axons release neurotransmitters that are probably glutamate and an eleven amino acid polypeptide called substance P. These trigger cord nerve cells to send signals along cord axons, which finally reach the somatosensory cortex in the brain by several ascending pathways. Signals from the brain down the spinal cord can then modulate the effect of the nociceptor. Descending axons end at the spinal levels on interneurons in the pain pathways as well as on afferent nociceptor neurons themselves. Some neurons in these inhibitory pathways release narcotic painkillers, *i.e.* endogenous opioids such as endorphins (endogenous morphine) and enkephalins that bind to various subtypes of opioid receptors (in the brain as well as the spinal cord) and mimic the effects of opium. All potent painkillers (morphine (6), codeine (7), *etc.*) act through these opioid receptors. The binding of an analgesic to μ receptors (one of a number of analgesic receptors) releases K$^+$ ions from

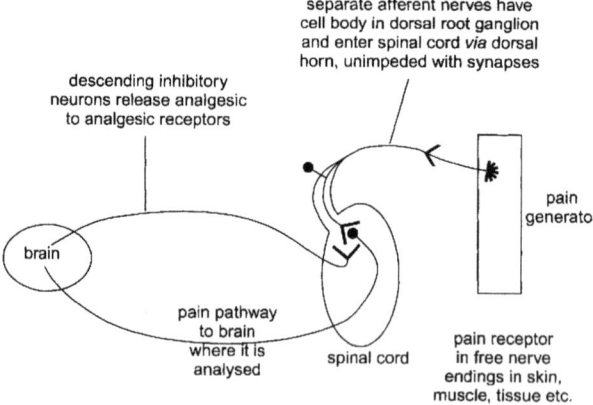

Figure 7.11 *Simplified signalling pathway for pain*

the cell leading to hyperpolarization and repression of the action potential. This also results in a decrease in Ca^{2+} ions entering the cell and reduced transmitter release. Both effects produce a shut down of the nerve and the blocking of pain messages.

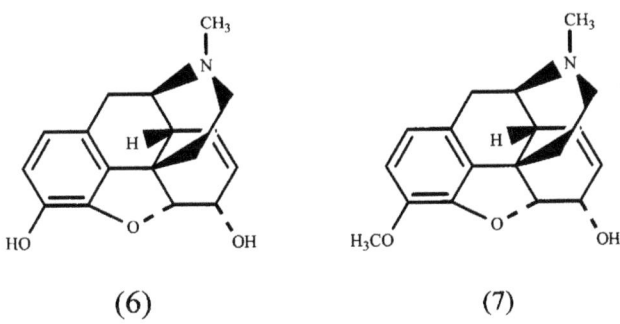

(6)   (7)

(6)(7) (−)Morphine (6) is the most important and powerful opium alkaloid. Derivatives are the diacetic ester (heroin) and the monomethyl ester (7) (codeine)

Block a Calcium Channel and Feel no Pain

It is known that several voltage dependent calcium channel (VDCC) blockers inhibit pain transmission. 'Knockout' mice that have been bred to be deficient in an N-type VDCC (which is neuron specific (5.1)) have suppressed responses to various pain stimuli that were sufficient to cause inflammation. A synthetic 25-residue peptide based on the cone snail toxin ω-conotoxin (Table 4.6) is an antagonist of N-type VDCC calcium channels that promote neurotransmitter release (5.1). The peptide prevents the propagation of pain signals, and in minute amounts is found to be a powerful analgesic and neuroprotective drug (Ziconotide). Controlling the activity of this channel might therefore be an approach to the alleviation of neuropathic pain in the future.

Muscle ischemia, inflammation and infection all invoke pain and are accompanied by local acidosis (an excess acid condition). An acid-sensing ion channel has recently been cloned. It is expressed in sensory neurons in dorsal root ganglia and is widely distributed in the brain. The acid-sensitive ion channel is closed at pH 7.4 and opens when the pH is <7.

7.7.2 Hot and Cold Pain

Extreme temperatures and noxious compounds can use the same receptor (Figure 7.12). We have seen that capsaicin activates a receptor (vanilloid,

VR1), which is mainly a Ca^{2+} ion channel. This allows movement of Ca^{2+} ions from sera ($[Ca^{2+}] = 2$ mM) into the intracellular space (initially, $[Ca^{2+}] = 100$ nM) and causes depolarization of pain neuronal fibres. The resulting pain signal proceeds from tongue to brain *via* the dorsal roots of the spinal cord. This same VR1 channel is also activated by an increase in temperature (*e.g.* from 22 °C to 43 °C, where heat becomes painful) and is therefore in addition a heat-sensitive receptor thus giving chilli peppers their heat sensation. Using genetically altered mice that lack the VR1 receptor supports this concept of dual response. VR1-null mice show much less long-term reaction (after the initial response) than do normal mice when their paws are injected with capsaicin or are placed on a hot plate. The VR1 knockout mice are otherwise normal. This is an encouraging step in the search for drugs that stop pain, but have few side effects. With prolonged exposure to capsaicin a Ca^{2+} ion overload is observed causing cell death and the nociceptor terminals become insensitive to capsaicin. Applied in a cream, capsaicin can thus act as an effective analgesic against chronic painful neuropathies. VR1 is also activated by resiniferatoxin (8). It is an ultrapotent capsaicin analogue (both have common vanillyl moieties) present in the acrid milky sap of the cactus-like plant *Euphorbia resinifera*. The dried latex of the plant (euphorbiam) was used for the treatment of pain during the reign of Emperor Augustus of Rome. One of the most ancient drugs still in use, resiniferatoxin is presently undergoing trials for the relief of the pain experienced by patients suffering from diabetic neuropathy.

Another receptor, VRL-1, has been isolated from rat brain, which

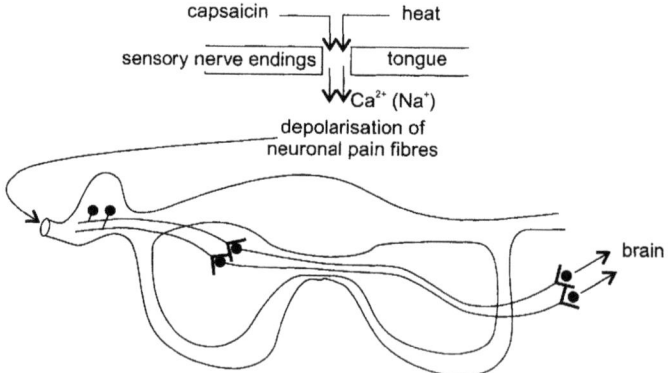

Figure 7.12 *Capsaicin and heat. This receptor has been cloned from the receptor neurons of rats and characterized. It opens for Ca^{2+} ions (visualized using a fluorescent indicator) in response to extreme temperatures and noxious compounds.*

responds to a greater temperature increase (~52 °C), but is not sensitive to capsaicin. It is still uncertain whether an endogenous ligand exists that is specific to VRL-1. Two candidates for the natural counterparts of capsaicin in the body are 12-HPETE (9) and anandamide (Sanskrit, *ananda*; bliss) (10). The structural similarities among them are apparent and all can activate vanilloid receptors.

Birds Can Act as Vectors for Seed Dispersal

Birds are immune to the hot taste of chilli peppers. Since, in addition, bird guts do not destroy chilli seeds there is a good chance they will be dispersed by the birds. An avian vanilloid receptor orthologue has been cloned from chicken sensory neurons. Although the channel still integrates noxious stimuli such as heat or protons, it has significantly reduced sensitivity to capsaicin. The sequences of the avian and mammalian capsaicin receptors have only 68% amino acid identity, which contributes to their different responses to vanilloid compounds.

For humans, a temperature of less than 15 °C is considered to be noxious cold. The CMR1 channel allows the detection of cold but not of ultracold (<8 °C) temperatures. This is in contrast to the VR1 and VR1L heat sensors that are activated only in the pain-producing temperature range. Other TRP channels (7.6) of which there are many in mammalian genomes, could modulate cold-induced pain.

7.8 SUMMARY

Changes in internal or external conditions are noted by an organism and then are acted upon. There are a number of special cell types that respond in the classical senses of touch, hearing, sight, smell and taste. Other sensations discussed include pressure, temperature and pain. These stimulations act on receptors that might incorporate ion channels or could give rise to second messengers that connect with ion channels. The production of receptor potentials and action potentials by the mechanisms detailed in Chapters 4 and 5 allow the transmission of signals (information) from the affected areas to the brain *via* specific fibres. The different sensory receptors employ a variety of mechanisms, ion channels and ions in order to function.

CHAPTER 8

Biomineralization

8.1 GENERAL FEATURES

Organisms ranging from bacteria and algae to vertebrates are capable of converting ions in solution to solid mineral salts. They achieve this with the aid of a relatively small number of biological macromolecules, which in the process become part of the whole array. The beautiful complex composite, the formation of which still defies complete understanding, has the desired property of hardness, but at the same time remains pliable. The deposition of inorganic salts in living organisms, called *biomineralization*, has impacts in fields as diverse as palaeontology, oceanography and limnology. The resistance of bone to decomposition is relied upon in palaeontology. In oceanography and limnology the effects of mineral sediments on, respectively, ocean and freshwater characteristics are important considerations. Over sixty different minerals are employed in biomineralization. Nearly thirty of these have calcium as their major cation, although iron and silicon are also well represented. Magnesium, strontium and barium ions have minor roles although magnesium is often a contaminant of calcium minerals, which are rarely deposited in a pure form (*e.g.* $MgCO_3$ is a contaminant of calcite in sponge spicules and strontium can substitute for calcium in aragonite in certain coral skeletons).

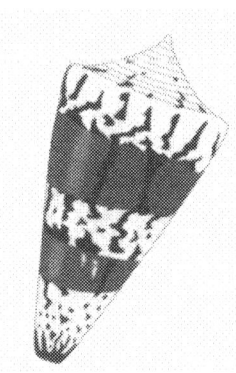

Magnesium/Calcium Ratios in Microfossil Shells Gives Some Insight into Past Climatic Conditions

Foraminifera are marine protozoa that secrete calcite. After they reproduce, the daughter cells abandon the parent shell and make new shells. The discarded shells cover millions of square kilometres of ocean floor to a depth of thousands of meters.

The Mg/Ca ratios (about 1 mmol mol^{-1}) in the shells of ocean-floor microfossils at different sites and depths (therefore of known age) have been measured by inductively coupled plasma–mass spectrometry (ICPMS) and atomic emission spectroscopy (ICPAES) (1.5.5). It is known that the partition coefficient of Mg^{2+} ions into calcite correlates well with temperature. In addition, tropical calcite shells are more enriched in magnesium than are those from subpolar regions. This fact, along with a good deal of subsidiary information, has allowed the Mg/Ca ratios in foraminifera to be used to estimate that deep sea temperatures have fallen about 12 °C over the past 50 million years. Each geological period has its own distinctive foraminiferan species and shell shapes (including chiral shells, 8.3.1) so the deposits are valuable age indicators.

Calcium ion usually occurs as carbonate and phosphate salts and a number of polymorphic forms of these are encountered. Calcium oxalates and sulfates are rare. The most important biominerals of Group 2 metal ions are shown in Table 8.1, which also gives examples of their occurrence.

The biomolecules taking part in biomineralization include small and large proteins (especially glycoproteins), carbohydrates and lipids. A mixture of these biomolecules is usually present and includes water-

Table 8.1 *Solid salts of Group 2 metal ions in biological systems*

Mineral	Occurrence
Calcite, $CaCO_3$	Widespread mineral in reptiles, birds and mammals as otoliths and egg shells.
	Common in invertebrate skeletons.
Aragonite, $CaCO_3$	Widespread, similar to calcite.
Hydroxyapatite, $Ca_{10}(PO_4)_6(OH)_2$	Prevalent in endoskeletons of vertebrates; *i.e.* teeth, bone (Ca store). Some in invertebrates.
Fluorapatite, $Ca_{10}(PO_4)_6(F)_2$	Shell in *Inarticulata* brachiopods.
	Articulata have calcite shells.
	Rare.
Fluorite, CaF_2	Rare; molluscs and chordata.
Weddellite, $CaC_2O_4.2H_2O$	Rare; molluscs and chordata.
MSO_4; M = Ca, Sr or Ba	Restricted occurrence.
	$CaSO_4$ a gravity device in jellyfish larvae.
	$SrSO_4$ a cellular support in *Acantharia* protozoa.
	$BaSO_4$ a gravity device in algae.
$CaMgP_2O_7$	Digestive glands of snails and slugs.
	Amorphous glanules occur inside a variety of cells (8.3.4).

soluble and water-insoluble fractions. They are often highly acidic proteins that contain many aspartic acid and glutamic acid residues. Therefore they are capable of metal ion coordination and this controls the important steps in the deposition process, namely nucleation, crystal orientation and polymorphic preference. The site of deposition of the mineral is often within an organism. Sometimes building materials are moved from inside the cell and the mineral ends up on the outer cell wall or epithelial surface. Different cell types can deposit different minerals even in the same organism. In the nacre-lined shells of abalone, for example, polyanionic proteins control the formation of calcium carbonate by using a genetic switch. The abalone first deposits a calcite layer and then abruptly changes to aragonite. It should be emphasized that most of the understanding of the control of crystal morphology and indeed of biomineralization in general is based on *in vitro* studies.

8.2 IMPORTANCE OF BIOMINERALS TO ORGANISMS

Skeletal support and protection are two paramount functions of biominerals. They are used in endoskeletons as supporting structures in vertebrates. Exoskeletons (external to the epidermis) are used for protection, as in the shells of molluscs, and are shed periodically.

Bigger and Better Antlers Led to the Demise of Irish Elk

Irish elk depicted in 22 000-year old cave paintings in France have enormous antlers. *Megaloceros* remains show that the antlers contain hydroxyapatite crystals. It has been estimated that the rack of an adult stag would contain 8 kg of calcium! During the 60-day period of antler mineralization each year, as much as 60 g of calcium would have to be deposited each day. Plants with the required calcium content probably disappeared around 10 000 years ago and Irish elk not long after. This may have lead to a seasonal osteoporosis that the animal could not sustain. It is surmised that female elk may have driven the males to evolve bigger and better antlers for show and weaponry and in the process this led to the extinction of the species.

A biomineral can also act as a buffer in an ion-storage system.

Turtles Use Their Shells During Anoxia

During prolonged periods of anoxia, oxygen does not reach the tissues of the freshwater turtle *Chrysemys picta belli*. In these circumstances its source of energy is derived from the conversion of carbohydrates to lactic acid. The extracellular buffer capacity in the turtle is unable to cope with the increased acid concentration. Using powdered turtle shells incubated in the physiologically relevant electrolyte solution, as well as from observations *in vivo*, it was shown that buffer capacity was enhanced by CO_2 (from carbonate) and by calcium and magnesium ions that are released *from* the shell. The amounts released depend directly on the acidity of the solution. The shell is also responsible for taking up lactate ions during anoxia. During normoxic recovery, the shell releases the required amount of lactate ion into circulation for reuse.

In some cases the biomineral is part of a buoyancy device.

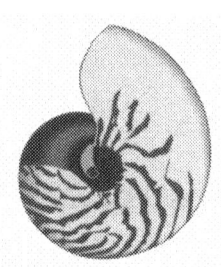

Its Shell Helps the Nautilus Adjust its Position in the Water

Pearly nautilus are beautiful Indo-Pacific sea creatures with many-chambered shells and numerous suckerless tentacles. The shell consists of layers of β-chitin (an *N*-acetylglucosamine polymer) and aragonite. Changes in the gas content in the shell chambers allow the creature to adjust its buoyancy and vertical position in the water. In addition the shell affords protection by allowing the body of the nautilus to withdraw into the last chamber of the tough hood. Nevertheless, boreholes have often been found in the shells of living animals that were drilled by octopus.

The major organisms in which biomineralization has been studied and is understood to varying extents will now be considered.

8.3 BIOMINERALIZATION IN INVERTEBRATES

8.3.1 Protoctista (Alternative, Protista)

Examples: protozoa, algae and dinoflagellates.

These simple eukaryotes are the main formers of sediment in the oceans. The shells and skeletons of the microscopic sea creatures, mainly foraminifera and coccolithophores, on the ocean floor as well as their dissolved products provide the buffer that maintains the pH of seawater at about 8.0. This is required since atmospheric carbon dioxide dissolving in surface seawater produces protons and bicarbonate ions. The 28 phyla include 19 mineral-forming species that use such diverse minerals as gypsum, $CaSO_4.2H_2O$, whewellite, $CaC_2O_4.H_2O$, barytes, $BaSO_4$ and cellestite, $SrSO_4$ as well as, primarily, calcite and silica (Table 8.1). The organic matrices employed include carbohydrates and their oxidation products (uronic acids) with the Ca^{2+} ions bound to carboxylate groups.

Taking advantage of the high reflectance of minerals in surface blooms of algae coccolithophorids in the North Atlantic, satellites have been able to detect, in a $7 \times 10^3 \, km^2$ area in the upper 60 m of the ocean, an estimated 7×10^4 tonnes of calcite. These algae form part of the oceanic phytoplankton. Their exquisite designs, ready availability, simple physiology and ease of handling have made them much favoured for the study of biomineralization.

Never is the manipulation of the inorganic component in biominerals by organic matter more exquisitely demonstrated than in the biological production of chiral shapes. Many coccolithophorids have their cell surfaces covered with coccoliths. Each coccolith consists of about 30 single calcite crystals arranged in an oval ring that exhibits macroscopic chirality.

Shell Coiling and Ocean Temperatures

The shells of the marine protozoa foraminifera *Globigerina pachyderma* (Ehrenberg) are made up of an inner organic layer and an outer calcite envelope. In the Arctic and Antarctic oceans more than 98% of the shells exhibit left coiling. In the temperate and tropical regions more than 98% of them coil to the right. Further, left coiling strongly dominates in sediment core samples one metre thick from top to bottom taken from the predominantly left coiling areas. In contrast, in core samples from the right coiling area of the North Atlantic there have been several reversals in the dominant direction that reflect climatic and sea-surface temperature changes. These observations indicate that there were southward shifts of isotherms in the North Atlantic during the last ice age. Nearly all the shells of gastropods (*e.g.* snails) coil only to the right. Very little is yet understood about these macroscopic chiral phenomena.

8.3.2 Cnidaria

Examples: jellyfish, sea anemones and corals (aquatic).

Stony coral polyps (the sedentary stage in the life cycle of the creature) secrete skeletal material from algae granules contained within them. Each polyp is connected to its neighbour by lateral strands. The interconnected polyps secrete an organic matrix upon which calcium carbonate (calcite that becomes aragonite in the adult) is deposited, the whole forming an exoskeleton. Here again, the organic matrix is complex, but contains calcium ion binding aspartate residues.

Dinoflagellates, a group of unicellular protoctista (8.3.1), live within the cells of corals, which affords them protection. In return the dinoflagellates photosynthesize carbohydrates for their host and aid in the biomineralization process in the coral. Reef-forming corals can only exist in shallow waters here light levels are sufficiently high to support photosynthesis and are warm enough for the organisms to survive. The extensive colonies thus consist of a living thin skin near the surface, which is supported by dead lace-like limestone where polyps once lived in the many holes.

The Great Barrier Reef of Australia, which can be seen from the moon, is over 2000 km long and up to 150 km wide and typically deposits 4 kg of calcium carbonate $m^{-2} yr^{-1}$. The largest limestone reserves worldwide consist mainly of coral remains. The Mg/Ca and Sr/Ca ratios in coral skeletons appear to be good indicators of sea-surface temperatures over the past two or three centuries (8.1). In certain coral species some of the variation in the Sr/Ca ratios can be ascribed to the presence of symbiotic algae (see above).

8.3.3 Crustacea

Examples: crabs, lobsters, shrimps and krill.

This subphylum is one of the largest groups in *Arthropoda*. These are creatures with a hard carapace or 'crust' that encloses the body. Tiny planktonic crustaceans in the Antarctic Ocean, called krill, are very abundant and are the main food of filter-feeding whales. The exterior of the carapace consists of proteins and glycoproteins, chitin and calcite (with small percentages of magnesium and phosphate ions). During moulting, old exoskeleton is decalcified prior to shedding and the new exoskeleton is mineralized soon after formation. This mineralization–demineralization cycle (using recycled mineral) and related aspects have been well studied. Membrane ion pumps (CaATPase and Na^+/Ca^{2+} exchanger) have important roles in ion accumulation in the biominerals.

8.3.4 Mollusca

Examples: snails, clams and squids.

Mollusca is a large and diverse phylum, the second largest on earth. Soft bodied, they are usually enclosed in a hard shell formed by a layer of tissue (*mantle*). Molluscs have well-developed sense organs and nervous systems (the squid has been a boon to neurophysiologists (4.3)) and are found in all in kinds of waters as well as on land.

Molluscs are the master builders of the natural world and epitomize biomineralization. As a consequence they have been well studied. They use a wide variety of minerals with calcium ions found as carbonate, phosphate, fluoride and oxalate salts in their shells and many other, mainly extracellular, structures as well. A single cell type can deposit more than one crystal form or mineral. This shell polymorphism is believed to be controlled by the organic matrix proteins that are secreted from the mantle epithelia. The soluble protein fraction contains many Ca^{2+}-binding aspartate and glutamate residues as well as ester sulfate groups. Nacre, or mother of pearl, forms the iridescent lining of many mollusc shells. This lustre arises from light interference patterns caused by alternating layers of aragonite and organic matter. Nacre is also the abnormal coating that is formed around a grain of sand in certain bivalve molluscs, which becomes a valuable gem, the pearl (from pearl oysters).

All Purpose Protein From Pearls

 Mollusc shells have two layers, a prismatic one containing calcite and a *nacreous* one with aragonite. A pearl inside the shell of a pearl oyster (*Pinctada fucata*) is equivalent to the nacreous shell layer. A 60 kDa protein, nacrein, is found in sizable amounts in the organic matrix extracted from powdered pearls. The protein was shown to have two functional domains:

- A domain with repeats of gly-asp (asn, glu)-asn residues that probably binds Ca^{2+} ions.
- A domain similar to that of the active site in carbonic anhydrases.

It is believed that the nacrein acts as a carbonic anhydrase and catalyses the conversion of respired CO_2 to HCO_3^-. It then binds Ca^{2+} ions and finally modulates a reaction with CO_3^{2-} ions to generate layers of aragonite.

There is a brick and mortar arrangement of the components in nacre. The organic mortar is a polysaccharide and protein fibre, which is similar to silk. This organic matrix, which is only about 1% by weight of the mineralized composite, is (perhaps surprisingly) the first portion of the sheets constructed by the cells. The matrix has many pores through which ions (and protein molecules) can flow to form the brick portion that consists of flat polygonal crystals of pure aragonite. This arrangement allows layers to align crystallographically *via* mineral bridges and to form the bricks at the required distance from the mantle cells, which are thus able to supply new materials for the growth of the nacre.

The transfer of ions from the external medium to finally form an insoluble coating on the shell is a complicated process and is still far from fully understood. The probable route taken by, and the ions involved in calcium carbonate shell formation, is shown in Figure 8.1.

There is no effect of microgravity (10^{-3}–10^{-4} G) on the shell building process in the freshwater snail *Biomphalaria glabrata*. Electron micro-

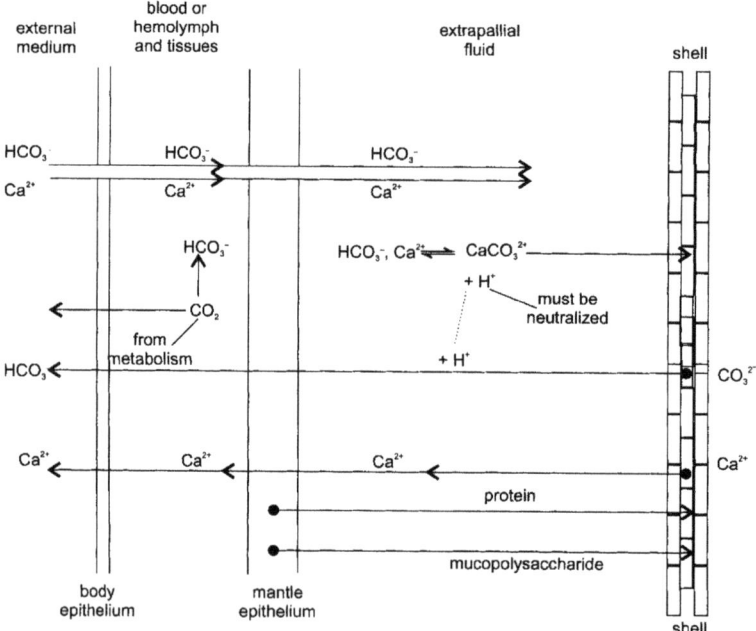

Figure 8.1 *The routes for calcium carbonate shell formation in molluscs. Calcium and bicarbonate ions must be moved towards the shell during deposition and away from it during its dissolution. This requires the input of energy, presumably supplied by ATP and ATPases. The bicarbonate ion, which ends up as carbonate in the shell, has been shown to originate from both the external medium and from tissue metabolism. There is extensive ion exchange at the body and mantle epithelia. The mantle secretes the organic materials used in the shell*

scopy (which employs an electron rather than a light beam) shows that the inorganic component, which makes up 99% of shell material, is itself 99% aragonite. It is built in the same regular manner in both adult snails and snail embryos. In the latter, shell building takes place even in the short duration (six days) of a space shuttle flight where the effect of no gravity could be assessed.

Mysterious Effect of Strontium on Statolith Growth

Statoliths are particles of dense calcareous material in the balance organs of molluscs. Octopus and squid that develop in synthetic ocean water containing no strontium were found to lack statoliths and moved with a spiral motion ('spinner syndrome'). Normal development of statoliths and motion occurred when ocean water containing 8 mg L^{-1} Sr (the usual concentration) was used during development. There is, as yet, no explanation for this strange phenomenon.

When a lightweight shell is required, either the mineral component is greatly reduced, as in land snails, or a very open matrix is used, as in the cuttlefish. Cuttlebone, the internal shell in cuttlefish, consists of layers of calcium carbonate spaced less than 1 mm apart. The layers are reinforced by chitin and supported by pillars of the same material. It is used by the cuttlefish to achieve neutral buoyancy (gas-filled) and is popular with bird fanciers as a source of calcium for their pets.

The inorganic material resulting from biomineralization is often non-crystalline. Inorganic granules occur inside a variety of cells, most commonly in invertebrates and particularly in molluscs. These are amorphous deposits about 100 μm in diameter and are located in membrane bound vesicles particularly those associated with the endoplasmic reticulum. The cations Ca^{2+} and Mg^{2+} and the anions CO$_3^{2-}$ and PO$_4^{3-}$ are those most commonly associated with these materials. All the granules also contain an organic matrix.

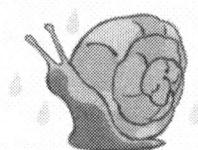

Snails Accumulate Toxic Metals

Inorganic granules occur most often in the digestive glands of molluscs, *e.g.* snails and slugs. These are also the sites for the accumulation of other metal ions. In the common garden snail (*Helix aspersa*) Ca$_2$P$_2$O$_7$ and

$Mg_2P_2O_7$ are the major components of the granules. The snail can concentrate cadmium and copper ions better than any other terrestrial invertebrate. Metallothionein is thought to be the ligand that transports these metal ions to the detoxification site. The site is in the hepatopancreas of snails and slugs where metal ions can be precipitated as pyrophosphates, for example, and where cells capable of producing amorphous intracellular granules are abundant.

8.4 BIOMINERALIZATION IN VERTEBRATES

Biomineralization has been considered only in invertebrates so far. A variety of biomineralization products found in living and fossil vertebrates will now be examined.

8.4.1 Otoliths

Otoliths (a general term) are dense stone deposits, precipitated in a protein matrix, found in the inner ears of vertebrates. They are present as either large ear stones (statoliths) or small particles of dust (statoconia). Bony fish (eels, salmon) have distinct statoliths while most other vertebrates have statoconia in an organic gel. Otoliths occur at the key sites for the sensing of balance and hearing (7.3). There are four forms of calcium carbonate and one of calcium phosphate found in the otoliths of vertebrates. They are:

- Calcite in reptiles, birds and mammals.
- Hydrated calcite in sharks and rays.
- Aragonite in fish (except sturgeon, which use vaterite).
- Poorly crystallized apatite in agnathans, primitive jawless vertebrates including lampreys and hagfish.

The otoliths of teleost (bony) fish consist of concentric rings that build up with age. Although the chemical composition of the rings is almost pure aragonite, they do contain contaminants, the amounts of which can be used to assess the environmental history of the fish. This is possible because otoliths, unlike other hard parts such as bone, are not subject to remodelling. The concentrations of strontium (one of the main contaminants, $SrCO_3$ has an aragonite-like structure) in the otoliths of marine habitat fish (*e.g.* mackerel, 0.24 wt.%) are 2–3 times greater than those of freshwater fish (*e.g.* mullet, 0.1 wt.%). This reflects the 10^2-fold higher concentration of Sr in seawater (8 ppm) than in most river water.

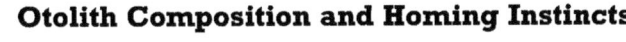

Otolith Composition and Homing Instincts

Salmon abandon salt water and swim up rivers to spawn in food-rich shallower waters. This has been well documented and is easily observed. Migration is much more difficult to investigate if the young fish are too small to tag and the fish spends its entire life in salt water.

The otoliths of juvenile weakfish, a blue-grey delicacy caught in natal areas in five major estuaries along the east coast of the US (from New York to Georgia) show significant differences in Mg/Ca, Mn/Ca, Sr/Ca and Ba/Ca compositions, thus fingerprinting each area. Two-year old adult weakfish ready to spawn in these areas have been caught and their otoliths examined. The compositions of the central cores generally matched well with those from the whole otoliths of the juvenile weakfish in each of the five estuaries collected two years previously. Most (60–80%) adults therefore return home to spawn. Unexpectedly, therefore weakfish are not a single coastwide family and stocking one area might not help another that has been overfished.

Not all Eels Have a Freshwater Growth Stage

An analysis of otoliths from a number of eel specimens taken from a river (the Elbe) and the sea (North Sea) was carried out using synchrotron radiation induced X-ray fluorescence imaging. Stones collected from the freshwater eels contained high [Sr^{2+}] in the core of the otoliths. This corresponded to larval migration from the spawning ground (in the sea) to the juvenile rearing habitat. The central core was surrounded by stone that had a much lower [Sr^{2+}]/[Ca^{2+}] ratio, corresponding to the entry of the juvenile eels into fresh water. This low ratio persisted until maturity. All eel specimens collected from the sea, however, had the higher [Sr^{2+}]/[Ca^{2+}] ratio *throughout the stone*, indicating that they had no freshwater history.

Otolith composition for diagnostic purposes requires careful analyses and increasingly sophisticated methods are being employed (1.5.5). The convening of international symposia on the subject and increased monetary support indicates the growing value of research into fish otoliths.

8.4.2 Bone

Bone is a highly structured porous material consisting of about 65% mineral (hydroxyapatite) interspersed with blood vessels and several proteins of which the major component is collagen fibres. Collagen is a common fibrous protein found in all multicellular animals, which is composed mainly of glycine and proline residues. Bone is the skeletal material in many animals. About 10% of skeletal bone mass is replaced annually in adult vertebrates. Bones are the major calcium and phosphate reservoirs in humans. This should not be surprising since 99% of the total calcium of the human body is contained in its bones. Hormones and vitamins play a major role in the formation and resorption of bone. The vitamin K dependent protein of bone, ostecalcin, which contains three gla residues, is one of the most abundant proteins in the human body. It is produced in osteoblasts (Figure 8.2) and its appearance in bone coincides with the onset of mineralization. The function of osteocalcin is unknown. It also occurs, together with other matrix gla proteins, in the cartilage and dentin of all vertebrates. The parathyroid gland releases parathyroid hormone (PTH) when plasma $[Ca^{2+}]$ is low. PTH stimulates osteoclasts to resorb bone and pulls calcium ions from bone to restore the normal Ca^{2+} ion concentration in serum. In the case of high blood $[Ca^{2+}]$, the thyroid gland secretes calcitonin, which stimulates osteoblasts to take up Ca^{2+} and deposit new bone. Over many years, however, bone density becomes depleted (*osteoporosis*).

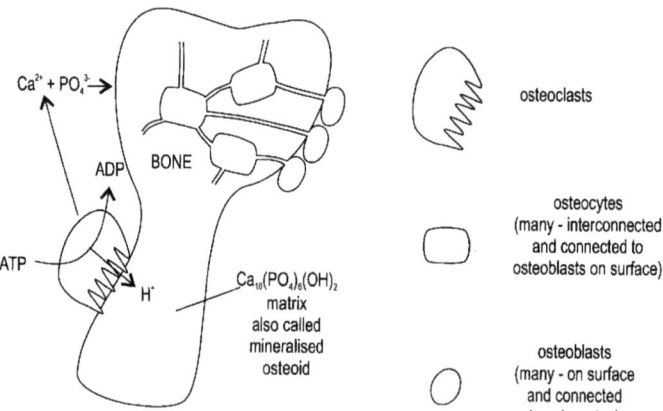

Figure 8.2 *Bone cells. There are three major types of bone cells embedded in the bone matrix: osteoblasts – secrete the bone matrix; osteocytes – these emerge from osteoblasts, perform cellular duties, maintain bone minerals and control exchange with plasma; osteoclasts – are large multinuclear cells involved in the breakdown of the bone matrix. They have ruffled surfaces that are the sites of bone resorption. The mobilization of Ca^{2+} ions is via the secretion of acids, chelating agents and extracellular proteases acting on the bone surface*

The malfunctioning of bone cells can have serious consequences.

Remember Cod Liver Oil?

Rickets is a disease characterized by low calcium and phosphate concentrations. The secretion of the osteoid matrix by osteoblasts is uncalcified and this causes deficient mineralization of bone. The result is soft pliable bones and crippling deformities. *Osteomalacia*, the adult equivalent of rickets, arises from the demineralization of preexisting bones that therefore fracture easily. Both conditions respond well to ingestion of cod liver oil (or simply a proper diet). The vitamin D group is contained in the fish oil and is the beneficial component. The fat-soluble steroid prohormone (in the oil) is hydroxylated in the liver and kidneys to produce calcitriol (1,25-dihydroxyvitamin D_3). Calcitrol promotes gut absorption and kidney resorption of calcium and phosphate ions, thus increasing their concentrations in the blood. In this way the important ingredients for good bone (and tooth) structure are restored.

Loss of the osteoid matrix and protein components causes brittle bones, leading to the possibility of hip fractures (*osteoporosis*). Postmenopausal osteoporosis is in some cases due to a genetic mutation.

In addition to hormone therapy for osteoporosis, vitamin D (above) and methylenebiphosphonates (MBP (1)), are increasingly popular, *e.g.* Fosamax, alendronic acid, $R = (CH_2)_3NH_2$.

(1)   (2)

MBPs resemble pyrophosphates (2), but are more stable to enzyme catalysed hydrolysis. They therefore bind strongly to calcium at the surface of hydroxyapatite crystals in growing or regenerating bone and thereby (somehow) inhibit bone resorption.

8.4.3 Teeth

Teeth are the hard bony outgrowths from the jaws of most vertebrates, which are used not only for biting and chewing, but also for attack! In mammals teeth consist of:

(a) A core of soft tissue (pulp) that is supplied by nerve and blood vessels.
(b) A layer of dentine, which has the same constituents as bone and forms the bulk of teeth.
(c) An outer covering of enamel, a hard material over 90% of which is comprised of calcium and magnesium phosphates.

Fluoride substitution for hydroxy groups in hydroxyapatite makes the material in the tooth less soluble and less sensitive to acid. This is a reason for inclusion of fluoride in toothpaste and water supplies.

Dentine is called ivory when it forms the major part of the tusks of elephants, walrus and other animals.

Pearly Teeth!

The materials in teeth and nacre are not homologous since they contain quite different calcium minerals, hydroxyapatite in teeth and aragonite in nacre. However teeth made of nacre, with roots that fit perfectly into the surrounding bone, have been discovered in the skulls of an ancient Indian people (Mayan) that inhabited parts of Central America.

Scientists have injected a slurry of a patient's blood and finely powdered nacre (from the mollusc *Pinctada maxima*) into tissues where the bone in the upper jaw was missing. Amazingly the nacre was not rejected and the bone-forming osteoblast cells were activated. New healthy bone was formed and the nacre was slowly lost. Eight female patients aged from 48–55 years old, suffering from bone loss in their upper jaws, were treated with this remarkable implantation method.

Calcification can sometimes occur where it shouldn't! A considerable number of diseases are associated with extraskeletal calcification. *Arteriosclerosis* arises, for example, from deposits of calcium salts in arteries resulting in a diminished blood supply especially to the brain and legs. A test that measures calcium deposits could be useful for assessing heart disease. Abnormal stones (*calculus*) form when mineral salts clump together anywhere in the body, but they are principally associated with the kidneys, bladder and joints. Kidney stones arise from the precipitation of salts (usually containing Ca^{2+} ions) from supersaturated urine. A number of forms of hypercalciuria (excess calcium ion in the urine) are, still inexplicably, due to mutations in the renal chloride ion channel CLCN5.

Crystal seeding centres are needed for precipitation and crystallization of inorganic salts even in supersaturated solutions. It has been suggested (controversially) that a newly discovered class of bacteria, called *nanobacteria* (50–500 nm in diameter), act as such nidi for the formation of *biogenic carbonate apatite* $(Ca_{10}(PO_4)_{6-x}(CO_3)_x(F,OH)_{2+x})$ on their cell envelopes at neutral pH. The precipitated salt could act as a protective coating. Certain bacteria in aqueous sediments are known to release oligopeptides that nucleate apatite.

A variety of diverse conglomerates of organic templates and calcium minerals are produced by living organisms. The results are shells and bones with outstanding properties closely related to their functions. As soon as (!) the principles behind the complex processes are fully understood, it is hoped that useful materials can be synthesized. These will mimic the superior properties (*e.g.* in strength or pliability) shown by the biological materials, for example, bone analogues for tissue engineering.

8.5 SUMMARY

In biomineralization calcium ions play a major role. This is one of the most beautiful and complex phenomena that is encountered in biology, but it is still poorly understood. It involves the conversion of ions in solution into solid minerals, both in vertebrates and invertebrates. It requires the assistance of (surprisingly sometimes) small amounts of organic material to provide the necessary combination of hardness and pliability. As well as the obvious skeletal role, biominerals are also important storage centres. The deposition of mineral is cell-specific. Nearly 30 different calcium containing minerals have been described so far. Very small contamination of some biominerals by magnesium and strontium can be used to give an insight into past climatic conditions and fish migratory patterns.

References

GENERAL TEXTS

There are many excellent general texts in biochemistry, biology and physiology. Some that we have relied upon heavily in writing this book include:

T. Nogrady, *Medicinal Chemistry*, 2nd Ed., OUP, New York, 1988.

H. Lüllman, K. Mohr, A. Ziegler and D. Bieger, *Pocket Atlas of Pharmacology*, Thieme, Stuttgart, 1993.

G. M. Shepherd, *Neurobiology*, 3rd Ed., OUP, New York, 1994.

D. Voet, J. G. Voet and C. W. Pratt, *Fundamentals of Biochemistry*, J. Wiley, New York, 1999.

C. Tudge, *The Variety of Life*, OUP, Oxford, 2000.

W. K. Purves, G. H. Orians and H. C. Heller, *Life – The Science of Biology*, 4th Ed., Sinauer Associates, Inc., Sunderland, MA, 1995.

E. Lawrence, *Henderson's Dictionary of Biological Terms*, 10th Ed., Longman, Harlow, England, 1989.

A. Vander, J. Sherman and D. Luciano, *Human Physiology*, 7th Ed., McGraw-Hill, Boston, 1998.

K. Schmidt-Nielsen, *Animal Physiology*, 5th Ed., Cambridge University Press, Cambridge, 1997.

D. Randall, W. Burggren and K. French, *Eckert Animal Physiology*, 4th Ed., W. H. Freeman, New York, 1997.

N. Sperelakis, Ed., *Cell Physiology Source Book*, 2nd Ed., Academic Press, San Diego, CA, 1998.

C. Hansch, P. G. Samnes and J. B. Taylor, Eds, *The Rational Design, Mechanistic Study and Therapeutic Application of Chemical Compounds*; Comprehensive Medicinal Chemistry in 6 Volumes, Pergamon Press, Oxford, 1990.

The Merck Index, An Encyclopedia of Chemicals, Drugs and Biologicals, 13th Ed., Merck and Co, Inc, Whitehouse Station, NJ, 2001.

CHAPTER 1 THE IONS

B. Crystall, "Monstrous Mucus", *New Scientist*, 2000, 11 March, 38–41.

C. M. Slupsky and B. D. Sykes, "The Structural Basis of Regulation by Calcium-Binding EF-Hand Proteins", in *Calcium as a Cellular Regulator*, E. Carafoli and C. Klee, Eds, OUP, New York, 1999, 73–99.

J. Rizo and T. C. Südhof, "C_2-domains, Structure and Function of a Universal Ca^{2+}-binding Domain", *J. Biol. Chem.*, 1998, **273**, 15879–15882.

S. R. Khan, Ed., *Calcium Oxalate in Biological Systems*, CRC Press, Boca Raton, FL, 1995.

R. Y. Tsien, "Fluorescence Imaging Creates a Window on the Cell", *Chem. Eng. News*, 1994, **72**, 18 July, 34–44.

A. Takahashi, P. Camacho, J. D. Lechleiter and B. Herman, "Measurement of Intracellular Calcium", *Physiol. Rev.*, 1999, **79**, 1089–1125.

Y. Rosenthal, M. P. Field and R. M. Sherrell, "Precise Determination of Element/Calcium Ratios in Calcareous Samples Using Sector Field Inductively Coupled Plasma Mass Spectrometry", *Anal. Chem.*, 1999, **71**, 3248–3253.

CHAPTER 2 BIOLOGICAL ROLES

S. R. Smedley and T. A. Eisner, "Sodium Uptake by 'Puddling' in a Moth", *Science*, 1995, **270**, 1816–1818.

C. L. Sonnichsen, *The El Paso Salt War (1877)*, Texas Western Press, El Paso, TX, 1961.

S. A. Lewis, "Epithelial Structure and Function in Epithelial Transport", in *Epithelial Transport. A Guide to Methods and Experimental Analysis*, N. K. Wills, L. Reuss and S. A. Lewis, Eds, Chapman and Hall, London, 1996, 1–20.

H. Sigel and A. Sigel, Eds., "Compendium on Magnesium and its Role in Biology, Nutrition and Physiology", *Metal Ions in Biological Systems*, Vol. 26, Dekker, New York, 1990.

E. Carafoli and C. Klee, Eds, *Calcium as a Cellular Regulator*, OUP, New York, 1999.

R. J. P. Williams, "Calcium: The Developing Role of its Chemistry in Biological Evolution" in *Calcium as a Cellular Regulator*, E. Carafoli and C. Klee, Eds., OUP, New York, 1999, 3–27.

C. L. Drum, S.-Z. Yan, J. Bard, Y.-Q. Shen, S. Soelaiman, Z. Grabarek, A. Bohm and W.-J. Tang, "Structural Basis for the Activation of Anthrax Adenyl Cyclase Exotoxin by Calmodulin", *Nature*, 2002, **415**, 396–402.

T. Drakenberg, S. Forsén and J. Stenflo, "Gamma-Glutamic Acid-Containing Proteins", in *Calcium as a Cellular Regulator*, E. Carafoli and C. Klee, Eds, OUP, New York, 1999, 123–151.

C. Paradiso, *Fluids and Electrolytes*, 2nd Ed., Lippincott, Philadelphia, 1999.

J. Schulkin, *Sodium Hunger – the Search for a Salty Taste*, Cambridge University Press, Cambridge, 1991.

J. Schulkin, *Calcium Hunger – Behavioural and Biological Regulation*, Cambridge University Press, Cambridge, 2001.

N. J. Birch, 'Inorganic Pharmacology of Lithium', *Chem. Rev.*, 1999, **99**, 2659–2682.

CHAPTER 3   MOVING IONS THROUGH MEMBRANES

C. I. Bargmann, "Neurobiology of the *Caenorhabditis elegans* Genome", *Science*, 1998, **282**, 2028–2033.

B. Hille, *Ion Channels of Excitable Membranes*, 3rd Ed., Sinauer Associates, Inc., Sunderland, MA, 2001.

D. J. Aidley and P. R. Stanfield, *Ion Channels*, Cambridge University Press, Cambridge, 1996.

W. A. Catterall, "From Ionic Currents to Molecular Mechanisms: The Structure and Function of Voltage Gated Sodium Channels", *Neuron*, 2000, **26**, 13–25.

M. L. Garcia, M. Hanner, H.-G. Knaus, R. Koch, W. Schmalhofer, R. S. Slaughter and G. J. Kaczovowski, "Pharmacology of Potassium Channels", *Adv. Pharm.*, 1997, **39**, 425–471.

R. W. Tsien and D. B. Wheeler, "Voltage-Gated Calcium Channels", in *Calcium as a Cellular Regulator*, E. Carafoli and C. Klee, Eds, OUP, New York, 1999, 171–199.

R. J. Miller, "Rocking and Rolling with Calcium Channels", *Trends in Neurosciences*, 2001, **24**, 445–449.

Y. Zhou, J. H. Morales-Cabral, A. Kaufman and R. MacKinnon, "Chemistry of Ion Coordination and Hydration Revealed by a K^+ Channel–Fab Complex at 2.0 Resolution", *Nature* 2001, **414**, 43–48.

F. Hucho and C. Weise, "Ligand-Gated Ion Channels", *Angew. Chem. Int. Ed., Eng.*, 2001, **40**, 3100–3116.

A. Miyazawa, Y. Fujiyoshi, M. Stowell and N. Unwin, "Nicotinic Acetylcholine Receptor at 4.6 Å Resolution: Transverse Tunnels in the Channel Wall", *J. Mol. Biol.* 1999, **288**, 765–786.

E. Carlsson, P. Lindberg and S. von Unge, "Two of a Kind", *Chemistry in Britain*, 2002, **38**, May, 42–45.

M. P. Blaustein and W. J. Lederer, "Sodium/Calcium Exchange: Its Physiological Implications", *Physiol. Rev.*, 1999, **79**, 763–854.

J. Orlowski and S. Grinstein, "Na$^+$/H$^+$ Exchangers of Mammalian Cells", *J. Biol. Chem.*, 1997, **272**, 22373–22376.

L. Fliegel, *The Na$^+$/H$^+$ Exchanger*, R. G. Landes Co., Austin, TX, 1996.

CHAPTER 4   INTRACELLULAR SIGNALLING

D. J. Aidley, *The Physiology of Excitable Cells*, 4th Ed., Cambridge University Press, 1998.

J. G. Nicholls, A. R. Martin, B. G. Wallace and P. A. Fuchs, *From Neuron to Brain*, 4th Ed., Sinauer Assoc., Sunderland, MA, 2001.

I. B. Levitan and L. K. Kaczmarek, *The Neuron – Cell and Molecular Biology*, 2nd Ed., OUP, Oxford, 1997.

R. R. Llinás, *The Squid Giant Synapse – A Model for Chemical Transmission*, OUP, New York, 1999.

E. Neher and B. Sakmann, "The Patch Clamp Technique", *Sci. Amer.*, 1992, **266**, March, 44–51.

A. L. Gotter, M. A. Kaetzel and J. R. Dedman, "Electrocytes of Electric Fish", in *Cell Physiology Source Book*, N. Sperelakis, Ed., 2nd Ed., Academic Press, San Diego, CA, 1998, Chapter 59.

F. M. Ashcroft, *Ion Channels and Disease*, Academic Press, San Diego, CA, 2000.

E. C. Cooper and L. Y. Jan, "Ion Channel Genes and Human Neurological Disease: Recent Progress, Prospects and Challenges", *Proc. Natl. Acad. Sci. USA*, 1999, **96**, 4759–4766.

R. A. Caras, *Dangerous to Man*, Chilton Books, Philadelphia, 1964.

F. Lehmann-Horn and K. Jurkat-Rott, "Voltage-Gated Ion Channels and Hereditary Diseases", *Physiol. Rev.*, 1999, **79**, 1317–1372.

W. T. Shier and D. Mebs, Eds, *Handbook of Toxinology*, Marcel Dekker, New York, 1990.

M. E. Adams and G. Swanson, "Neurotoxins", 2nd Ed., in *Trends Neurosci. Supplement*, **19**, June, 1996.

B. Furlow, "The Freelance Poisoner", *New Scientist*, 2001, **170**, 20 Jan., 30–33.

T. Yasumoto and M. Murata, "Marine Toxins", *Chem. Rev.*, 1993, **93**, 1897–1909.

P. Berressem, "From Bites and Stings to Medicines", *Chemistry In Britain*, 1999, **35**, 40–42.

CHAPTER 5   INTERCELLULAR SIGNALLING

W. M. Cowan, T. C. Südhof, C. F. Stevens and K. Davies, Eds, *Synapses*, The Johns Hopkins University Press, Baltimore, MD, 2001.

J. Stephen and R. A. Pietrowski, Eds, *Bacterial Toxins*, 2nd Ed., American Society for Microbiology, Washington, DC, 1986.

G. Schiavo, M. Matteoli and C. Montecucco, "Neurotoxins Affecting Neuroexocytosis", *Physiol. Rev.*, 2000, **80**, 717–766.

W. N. Arnold, "Absinthe", *Sci. Amer.*, 1989, **260**, June, 112–117.

D. Barchan, S. Kachalsky, D. Neumann, Z. Vogel, M. Ovadia, E. Kochva and S. Fuchs, "How the Mongoose Can Fight the Snake: The Binding Site of the Mongoose Acetylcholine Receptor", *Proc. Natl. Acad. Sci. USA*, 1992, **89**, 7717–7721.

CHAPTER 6   MUSCLE

A. J. McComas, *Skeletal Muscle. Form and Function*, Human Kinetics, Champaign, IL, 1996.

S. V. Perry, *Molecular Mechanisms in Striated Muscle*, Cambridge University Press, Cambridge, 1996.

R. H. Adrian and S. H. Bryant, "On the Repetitive Discharge in Myotonic Muscle Fibres", *J. Physiol.*, 1974, **240**, 505–515.

M. W. Berchtold, H. Brinkmeier and M. Müntener, "Calcium Ion in Skeletal Muscle: Its Crucial Role for Muscle Function, Plasticity and Disease", *Physiol. Rev.*, 2000, **80**, 1215–1265.

S. Ebashi, M. Endo and I. Ohtsuki, "Calcium in Muscle Contraction", in *Calcium as a Cellular Regulator*, E. Carafoli and C. Klee, Eds, OUP, New York, 1999, 579–595.

G. H. Pollack, *Cells, Gels and the Engines of Life*, Ebner and Sons, Seattle, WA, 2001.

I. Rayment and H. M. Holden, "The Three-Dimensional Structure of a Molecular Motor", *Trends Biol. Sci.*, 1994, **19**, 129–134.

L. C. Rome, D. A. Syme, S. Hollingworth, S. L. Lindstedt and S. M. Baylor, "The Whistle and the Rattle: The Design of Sound Producing Muscles", *Proc. Natl. Acad. Sci. USA*, 1996, **93**, 8095–8100.

R. R. Swaisgood, M. P. Rowe and D. H. Owings, "Assessment of Rattlesnake Dangerousness by California Ground Squirrels: Exploitation of Cues From Rattling Sounds", *Anim. Behav.*, 1999, **57**, 1301–1310.

A. M. Katz, *Physiology of the Heart*, 2nd Ed., Raven Press, New York, 1992.

H. Brown and R. Kozlowski, *Physiology and Pharmacology of the Heart*, Blackwell Science, Oxford, 1997.

"Nature Insight – The Heart", *Nature*, 2002, **415**, 197–243.

W. G. Naylor, *Calcium Antagonists*, Academic Press, London, 1988.

CHAPTER 7   SENSES

"Nature Insight – Molecular Sensing", *Nature*, 2001, **413**, 185–230.

E. B. Levitan and L. K. Kaczmarck, "Sensory Receptor Neurons", in *The Neuron – Cell and Molecular Biology*, 2nd Ed., OUP, Oxford, 1997, Chapter 13.

T. E. Finger, W. L. Silver and D. Restrepo, Eds, *The Neurobiology of Taste and Smell*, 2nd Ed., Wiley-Liss, New York, 2000.

D. J. Aidley, "Sensory Cells" in *The Physiology of Excitable Cells*, D. J. Aidley, Ed., 4th Ed., Cambridge University Press, 1998, Chapters 13–17.

R. Axel, "The Molecular Logic of Smell", *Sci. Amer.*, 1995, **273**, Oct., 154–159.

D. Schild and D. Restrepo, "Transduction Mechanisms in Vertebrate Olfactory Receptor Cells", *Physiol. Rev.*, 1998, **78**, 429–466.

D. V. Smith and R. F. Margolskee, "Making Sense of Taste", *Sci. Amer.*, 2001, **284**, March, 29–32.

D. M. McKemy, W. M. Neuhausser and D. Julius, "Identification of a Cold Receptor Reveals a General Role for TRP Channels in Thermosensation", *Nature*, 2002, **416**, 52–58.

CHAPTER 8   BIOMINERALIZATION

K. Simkiss and K. M. Wilbur, *Biomineralization. Cell Biology and Mineral Deposition*, Academic Press, San Diego, CA, 1989.

L. Addadi and S. Weiner, "Control and Design Principles in Biological Mineralization", *Angew. Chem. Int. Ed. Engl.*, 1992, **31**, 153–169.

C. H. Lear, H. Elderfield and P. A. Wilson, "Cenozoic Deep-Sea Temperatures and Global Ice Volumes from Mg/Ca in Benthic Foraminiferal Calcite", *Science*, 2000, **287**, 269–272.

R. T. Hanlon, J. P. Bidwell and R. Tait, "Strontium is Required for Statolith Development and the Normal Swimming Behaviour of Hatchling Cephalopods", *J. Expl. Biol.* 1989, **141**, 187–195.

D. Carlstrom, "Crystallographic Study of Vertebrate Otoliths", *Biol. Bull.*, 1963, **125**, 441–463.

G. Atlan, N. Balmain, S. Berland, B. Vidal and E. Lopez, "Reconstruction of Human Maxillary Defects With Nacre Powder: Historical Evidence for Bone Regeneration", *CR Acad. Sci.* III, 1997, **320**, 253–258.

T. Kee and N. Dixon, "Living Dangerously", *Chemistry in Britain*, Oct., 2001, 38-41.

S. Mann, *Biomineralization: Principles and Concepts in Bioinorganic Materials Chemistry*, OUP, Oxford, 2001.

Subject Index

omeprazole, 63
opsins, 164
organics in biominerals, 185–187
osteoblasts, 192–194
osteocalcin, 35, 192
osteoclasts, 192
osteocytes, 192
osteomalacia, 193
osteoporosis, 37, 183, 192, 193
otolith, 16, 163, 190, 191
oubain, 137
oxalate poisoning, 11
oxytocin, 102

P53, 33
pacemaker cells, 132, 138–140
pacemaker potential, 132
pain, 175–179
pancreatic cell, 52
papaverine, 142, 143
paramecium, 79, 158, 159
paramyosin, 145
paramyotonia congenita, 149, 151
parasympathetic nervous system, 140
parathyroid hormone, 192
parietal cell, 63, 64
paroxetine, 103, 104
parvalbumin, 28, 129
passive spread, 81
patch clamp, 76, 99
pawn mutant, 159
pearl, 187
pearl oysters, 187
pentalenene, 26
pentalenene synthase, 26
peppermint, 174
perineurium, 78
Peruvian Paso horses, 109
pH regulation, 65, 66
phenylalkylamines, 142
phenytoin, 87
philanthotoxin, 111, 112
phosphatidylinositol-4,5-biphosphate, 39, 59
phosphodiesterases, 25, 164, 166

phospholipase A_2, 8
phospholipase C, 8, 28, 32, 59, 101, 102, 166
phosphorylated MLCK, 143, 144, 146
photosynthesis, 26
picrotoxinin, 107, 108
pig, 149, 152
pit viper, 174
Plantago, 2
plant neurotoxins, 93
plasma membrane, 42
plasticity, 110
porpoise, 169
porcupine, 18
positive ionotropic effect, 137
postsynaptic axon, 99
postsynaptic potential, 104
potassium activated myotonia (PAM), 149, 150
potassium channels
    action potentials, 75, 131, 132
    agonists, 52
    antagonists, 52
    antiarrythmic drugs, 136
    characteristics, 45–47
    diabetes, 52
    functions, 51
    G-protein gated, 58, 140
    heartbeat, 140
    mutations, 52, 86, 163
    structural aspects, 4, 5, 48–52
    taste, 166, 171, 172
    toxins, 92, 93
    vascular smooth muscle, 146
potassium ions
    action potentials in cardiac muscle, 131, 132
    action potentials in squid, 75
    biological roles, 21
    cardioplegic solution, 2
    coordination geometry, 5, 8, 9, 51
    dietary requirements and problems, 37–39
    enzyme activation, 9, 21
    formation constants, 6